Shedding Light on Indoor Tanning

Carolyn J. Heckman, PhD • Sharon L. Manne, PhD
Editors

Shedding Light on Indoor Tanning

Springer

Editors
Carolyn J. Heckman
Cancer Prevention and Control Program
Fox Chase Cancer Center
333 Cottman Avenue, Rm P4163
Philadelphia, PA 19111
USA
Carolyn.Heckman@fccc.edu

Sharon L. Manne
The Cancer Institute of New Jersey
UMDNJ-Robert Wood Johnson
Medical School
195 Little Albany Street
New Brunswick, NJ 08901
USA
mannesl@umdnj.edu

ISBN 978-94-007-2047-3 e-ISBN 978-94-007-2048-0
DOI 10.1007/978-94-007-2048-0
Springer Dordrecht Heidelberg London New York

Library of Congress Control Number: 2011938106

Printed on acid-free paper

Springer is part of Springer Science+Business Media (www.springer.com)

Foreword

It is now clear that the use of tanning beds is associated with increased risk for squamous cell carcinoma, basal cell carcinoma and melanoma. In fact, in 2009, the World Health Organization categorized tanning beds as carcinogenic to humans [1]. Several countries worldwide have either banned the sale of tanning beds, or restrict access to tanning beds by minors; many states in the US have also limited access to minors. However, despite the known risk of exposure to tanning beds and the intense public education effort on the side-effects of tanning by many organizations in the US, tanning beds remain widely used, especially by young women.

Tanning beds are currently classified by the US Food and Drug Administration (FDA) as Class I medical devices with 510(k) exemption. This is the classification for devices that pose minimal potential harm, and are exempted from premarket notification to demonstrate their safety and efficacy. Aside from tanning beds, tongue depressors, bedpans, and elastic bandages are examples of Class I medical devices. Because of the public health concerns associated with tanning bed exposure, as evidenced by multiple peer reviewed publications, many professional medical organizations, including the American Cancer Society, American Academy of Dermatology, National Council on Skin Cancer Prevention, and others, have strongly urged the FDA to reevaluate the above classification. As a result, on March 25, 2010, the FDA conducted a public hearing on this issue. Topics presented included the effects of ultraviolet radiation (UV) exposure such as DNA damage and immune suppression, the observation that UVA output from tanning beds could be as high as four times noon sunlight in Washington, DC, the risk of skin cancer development and its economic burden, the vitamin D controversy, tanning as an addictive behavior, and the compliance of tanning bed operators to state and federal guidelines [2]. At the time of this writing, the final action of the FDA has not yet been released.

Drs. Heckman and Manne are to be congratulated for editing a book in which the relevant topics on indoor tanning are addressed in an evidenced-based manner by recognized leaders. Topics covered include the history and prevalence of tanning, health behavior theories on motivations for tanning, tanning addiction, the physiology of tanning, health effects of UV exposure (e.g., skin cancer development, vitamin D production), and intervention and regulatory efforts to minimize indoor tanning; the book concludes with discussion on sunless tanning as well as sunbed use outside of the US.

The timely publication of this book nicely fulfills the need of a single volume, up-to-date source of information on indoor tanning, a topic that has significant public health consequences. This book should be on the shelf of dermatologists who counsel patients daily on the side-effects of UV, policy makers who address the public health impact of indoor tanning, photobiologists and other scientists who investigate the cutaneous effects of UV radiation, and social scientists who study tanning motivations and design interventions to reduce engagement in the behavior.

Henry W. Lim, MD
Chairman and C.S. Livingood Chair
Department of Dermatology
Henry Ford Hospital, Detroit, Michigan, USA

References

1. El Ghissassi F, Baan R, Straif K et al (2009) A review of human carcinogens – part D: radiation. *Lancet Oncol* 10(8):751–752
2. Lim HW, James WD, Rigel DS et al (2011) Adverse effects of UV radiation from the use of indoor tanning equipment: time to ban the tan. *J Am Acad Dermatol* 64(4):e51–e60

Contents

Chapter 1
Introduction

Carolyn J. Heckman and Sharon L. Manne

Abstract Since the industrialization of the Western workforce, tanned skin has been perceived increasingly as attractive and fashionable for naturally light-skinned individuals. However, in addition to causing tanning, photo-aging, and other health effects, ultraviolet radiation (UV) is a well-known carcinogen. Despite increased awareness of UV risks, tanning has become widespread. While millions of individuals tan indoors each day, relatively little is known about this phenomenon. This book attempts to fill that gap by providing an overview of the phenomenon of indoor tanning, reasons for its popularity, its risks, and the public health context surrounding the behavior. We have invited some of the preeminent experts in the field to summarize the existing scientific literature for each of the chapters, which are described below. This book provides a unique and essential overview of the most significant current issues related to indoor tanning for scientists, educators, students, clinicians, and the general public interested in dermatology, aesthetic trends, skin care, and skin cancer.

Keywords Indoor tanning • Introduction • History of tanning • Prevalence of indoor tanning • Correlates of indoor tanning • Motivations for indoor tanning • How UV tans skin • Skin cancer • Tanning dependence • Vitamin D • Interventions to reduce indoor tanning • Regulation of indoor tanning • Sunless tanning • International issues in indoor tanning • Melanoma • Ultraviolet radiation (UV)

C.J. Heckman (✉)
Cancer Prevention and Control Program, Fox Chase Cancer Center,
333 Cottman Avenue, Philadelphia, PA 19111, USA
e-mail: carolyn.heckman@fccc.edu

S.L. Manne
The Cancer Institute of New Jersey,
UMDNJ-Robert Wood Johnson Medical School,
195 Little Albany Street, New Brunswick, NJ 08901, USA
e-mail: mannesl@umdnj.edu

C.J. Heckman and S.L. Manne (eds.), *Shedding Light on Indoor Tanning*,
DOI 10.1007/978-94-007-2048-0_1, © Springer Science+Business Media B.V. 2012

A tanned appearance has not been viewed favorably at all times in history, and some cultures favor a paler skin tone. Chapter 2 of the book provides a valuable historical and cultural perspective of tanning and its popularity. Which skin colors are most attractive is a subjective perception that has changed over time and across cultures. Additionally, the social meaning attributed to skin colors is mutable. This chapter outlines a fascinating history of tanning behavior and related attitudes and highlights the influential the role of fashion and advertising.

Is indoor tanning only a phenomenon of young, Caucasian women and girls, or is it more widespread? The current prevalence of indoor tanning and the demographic correlates predictive of who tans indoors, such as age, race, and gender, are described in Chap. 3. Prevalence rates of indoor tanning vary depending on the country and population under study as well as the time-frame being assessed. For example, the prevalence of indoor tanning in the past year among adults was highest in Europe, followed by the USA, and was much lower in Australia. Overall, indoor tanners are more likely than non-tanners to be female, adolescents or young adults, Caucasian, and to have low to moderate skin sensitivity to ultraviolet radiation (UV). Other correlates of indoor tanning include associating with other indoor tanners, as well as use of alcohol, cigarettes, and other substances.

Knowledge of the link between UV radiation exposure and skin cancer is widespread in the USA and several other Western countries. However, there are significant psychosocial motivations to tan that sometimes outweigh an individual's concern for his or her health. Chapter 4 describes the major theories explaining why people may tan indoors. Appearance enhancement is the most commonly-cited reason given for intentional indoor tanning. Attractiveness and the appearance of youth and vitality are highly prized in American culture. Indoor tanning is perceived to be an efficient and convenient way to tan, particularly in climates that are not conducive to continuous sun-tanning throughout the year. In addition to appearance enhancement, some individuals tan for mood enhancement or for social reasons. These can be very compelling factors, making individual behavior change and implementation of public health campaigns challenging.

Although many individuals find a tanned appearance attractive, it is also a sign of skin damage. UV radiation activates pigmentation in skin cells, producing a tanned appearance. A tanned appearance is a sign of underlying DNA damage to skin cells. In fact, UV radiation is a well-established human carcinogen. In addition to the psychosocial context of the behavior, it is important to understand the biological effects of UV radiation on the skin. Chapter 5 addresses how tanning and burning occur and how this varies by individual characteristics such as skin type and family heritage.

The major health risk of UV exposure is skin cancer, which is the most common cancer in the USA, with over a million new cases diagnosed yearly. Perhaps because tanning is commonly perceived as a cosmetic adolescent fad, its association with fatal skin cancer is not taken as seriously as it could be. Melanoma is the most lethal form of skin cancer, but non-melanomas can cause significant morbidity and even mortality as well. It is generally accepted that UV exposure is the most significant modifiable risk factor in the prevention of melanoma, and UV radiation has also been definitively demonstrated to cause non-melanoma skin cancers. In addition to increased risk for melanoma and non-melanoma skin cancers, UV radiation causes photo-aging (the visible signs of aging such as wrinkles and age spots) and can have negative effects on the immune system. Chapter 6 describes the health effects of UV radiation and indoor tanning.

There are also several current controversies that are addressed in the book including tanning addiction and myths perpetuated by the tanning industry. Appearance enhancement is the most commonly-cited reason given for intentional indoor tanning. An additional reason for frequent tanning, particularly indoor tanning, may be tanning dependence or addiction, colloquially referred to as "tanorexia". Chapter 7 describes the accumulating evidence, both observational and experimental, regarding the phenomenon of tanning dependence. Behavioral studies of adolescents and young adults have reported addictive tendencies among indoor tanners including higher rates of other substance use and anticipated difficulty quitting indoor tanning. Several studies have also found greater symptoms of other psychiatric disorders such as anxiety and mood problems among indoor tanners and tanning dependent individuals. Methods for defining and identifying tanning dependence are being refined, and a potential biological mechanism involving endogenous opioids has been proposed. The cutting-edge clinical experiments that have been conducted related to tanning dependence and its mechanisms are fascinating, even demonstrating altered brain activity during UV exposure. This topic is an excellent example of translational research linking basic and behavioral science.

Chapter 8 discusses some of the myths about tanning perpetuated by the tanning industry. One of the main reasons offered in defense of tanning by the tanning industry is the health benefit of vitamin D (e.g., bone health, colon cancer prevention), which is produced by the skin after UV exposure. No published studies have examined vitamin D levels in frequent tanners or tanning dependent individuals. While some level of vitamin D is undoubtedly beneficial for health, it is readily available as an oral supplement, and the high prevalence of vitamin D deficiency and claimed health benefits of high vitamin D levels are unproven. This is currently a very hot topic. The chapter also discusses the myths that indoor tanning is safer than sunbathing, that there is such a thing as a beneficial "base tan," and that commercial indoor tanning can be used as an alternative to medically-supervised phototherapy for skin conditions.

With the growing popularity of indoor tanning and its related consequences, a number of researchers have been attempting to develop and test interventions to reduce this behavior. Chapter 9 summarizes the research examining psychosocial interventions to reduce indoor tanning, particularly interventions that target appearance-related concerns. The literature is small but growing, and more rigorous intervention research is needed. This chapter outlines issues that should be addressed in future research on indoor tanning. For example, how can we combat the strong societal pressure, particularly among women, for appearance-enhancement by any means possible, regardless of the potential risks?

In addition to psychosocial interventions, policy interventions are also discussed. Chapter 10 describes the state and federal policies surrounding the regulation, enforcement, and taxation of the indoor tanning industry. Why are there more tanning salons in major American cities than McDonald's or Starbucks? Are there similarities between the tanning and tobacco industries with regard to marketing and regulation issues? Policy level interventions have been some of the most successful strategies for lowering smoking rates. This chapter describes the restrictions on indoor tanning primarily in the USA, some of which have occurred very recently and will likely further evolve in the future. These are restrictions related to age, parental consent, UV radiation levels, warning labels, taxation, and operator education. Unfortunately, these regulations have not had their intended impact due to insufficient compliance and enforcement. Successful tobacco regulation is used as a model to suggest strategies to reduce indoor tanning, particularly among our vulnerable youth.

Despite the book's primary focus on indoor UV tanning, Chap. 11 provides valuable information about chemical sunless tanning. What is sunless tanning and how does it work? Who sunless tans, and how does sunless tanning relate to UV tanning behavior? Sunless tanning is a safe alternative to UV tanning that can achieve the desired appearance-enhancement effects in an efficient and cost-effective manner. Sunless tanning has become quite popular in the USA in the last few years, and there are many varieties of sunless tanning products available. Chapter 11 will also address what interventions have been explored to alter sunless tanning behavior and what additional research is needed.

While the emphasis of this book and much of the existing research is on indoor tanning practices in the USA, Chap. 12 describes the prevalence, correlates, and regulation of indoor tanning in Europe, Canada, Australia, and other countries. The similarities and differences among the countries are fascinating and warrant additional research. Indoor tanning is truly a global issue.

In summary, indoor tanning is an all-too common behavior that is associated with multiple adverse effects, including skin cancer. Like tobacco, indoor tanning has been marketed to adolescents and young adults, putting this vulnerable population at increased risk for high levels of long-term UV exposure. Appearance enhancement is the primary motivation for indoor tanning, but there are several other potential reasons for the behavior, including regulation and enhancement of mood as well as tanning dependence. Recent regulations and interventions have been shown to have some effect on reducing indoor tanning rates, but there is considerable need for further research and education as well public health and policy efforts. The concluding chapter, Chap. 13, summarizes future directions for research, policy, and intervention.

Shedding Light on Indoor Tanning is a unique, up-to-date, and comprehensive book that we hope readers will find as interesting, informative, and useful as we do.

Acknowledgments The editors would like to thank Jeanne Pomenti for her extensive technical assistance in the preparation of this book. Drs. Heckman and Manne are funded by grants from the National Cancer Institute.

Chapter 2
History and Culture of Tanning in the United States

Yvonne Hunt, Erik Augustson, Lila Rutten, Richard Moser, and Amy Yaroch

Abstract This chapter traces changes in the perception of tanning and tanning behavior primarily within the United States (U.S.) from the later part of the nineteenth century to the early part of the twenty-first century. Originally seen as a hallmark of the working class/disadvantaged groups and associated with disease and ill health, societal perceptions of the tan evolved over time to reflect the opposite: wealth, health and beauty. These core beliefs regarding the value of tanning and ultraviolet (UV) radiation exposure have proven extremely difficult to modify despite substantial efforts by the public health community to do so. In an attempt to understand why millions of Americans continue to engage in high-risk, intentional UV exposure such as use of indoor tanning facilities, the beliefs and behaviors related to tanning are considered within the context of the historical medical and societal factors, especially the role of fashion and advertising, which helped to shape current opinion.

Keywords Advertising • History of tanning • Indoor tanning • Melanoma • Skin cancer • Sun bed • Sun exposure • Sun protection • Sun safety • Sunbathing • Sunburn • Sunlamp • Sunscreen • Tanning • Tanning bed • Ultraviolet radiation

---

Y. Hunt (✉) • E. Augustson • R. Moser
Behavioral Research Program, Division of Cancer Control and Population Sciences, National Cancer Institute, 6130 Executive Blvd, Ste 4042, Bethesda, MD 20892, USA
e-mail: huntym@mail.nih.gov; augustse@mail.nih.gov; moserr@mail.nih.gov

L. Rutten
Clinical Monitoring Research Program, SAIC-Frederick Inc., National Cancer Institute at Frederick, 5705 Industry Lane, Frederick, MD 21702, USA
e-mail: finneyl@mail.nih.gov

A. Yaroch
Gretchen Swanson Center for Nutrition, 505 Durham Research Plaza, Rm 1024, Omaha, NE 68105-1313, USA
e-mail: ayaroch@centerfornutrition.org

C.J. Heckman and S.L. Manne (eds.), *Shedding Light on Indoor Tanning*,
DOI 10.1007/978-94-007-2048-0_2,

Abbreviations

FDA	Food and Drug Administration
SPF	Sun Protection Factor
U.S.	United States
UV	Ultraviolet radiation
UVA	Ultraviolet A radiation
UVB	Ultraviolet B radiation
UVC	Ultraviolet C radiation

2.1 The Great American Tan

Ultraviolet (UV) radiation has deleterious effects on human health. Chronic UV exposure can result in permanent structural damage to the skin, including wrinkling, discoloration, and loss of elasticity. More importantly, UV radiation causes DNA damage, which can lead to the development of skin cancers [1–3].

Despite decades of public health efforts to communicate the importance of protection from UV radiation, millions of Americans continue to deliberately engage in behaviors that increase their UV exposure. As a result, the past 30 years have seen alarming upward trends in the incidence of skin cancers, including malignant melanoma, which is associated with poor survival rates [4–9]. A substantial body of published survey data suggests that most Americans are at least somewhat aware of the risks of excessive UV exposure [10–13]. Yet, cultural standards of attractiveness continue to place a high value on tanned skin at the expense of health and safety.

Why are so many Americans willing to risk their lives for a tan? In answer to this question, there is much to be learned by examining the historical and cultural context that has shaped the social construction of attitudes, beliefs, and behaviors around tanning. This information provides a framework for our understanding of current challenges related to skin cancer prevention. The image of a tan as healthy, attractive, and fashionable has proven difficult to reverse, and a well established tanning culture persists in the face of the known medical risks of skin cancer and photodamage. Public health professionals continue to be challenged by the need to address important questions about how to change the cultural norms around tanning in the U.S. In this chapter, we will begin with a discussion of some of the meanings associated with skin color and then consider how medical, social and economic factors have changed over time and what their impact on America's stance toward tanning has been.

2.2 Skin Deep

Skin color arises from a protein in the skin called melanin, which confers different shades of pigmentation when present in varying amounts [14, 15]. Melanin acts as a natural "sunscreen" by blocking the penetration of UV radiation through the skin.

Amounts of pigmentation differ around the world, with various populations demonstrating a wide range of tones from fair to dark, but most falling somewhere in between (i.e., brown) [16]. Geographical differences in skin pigmentation are hypothesized to be evolutionary adaptations to the UV light intensity in a particular population's native habitat [15]. Heavily pigmented populations are typically encountered in environments characterized by intense UV exposure, whereas fair-skinned populations are found in higher latitudes where the amount of solar radiation is less. In light-skinned humans, tanning is a natural response to the injurious effects of UV radiation [17]. The characteristic darkening of the skin (i.e., tanning) arises from the stimulation of melanin production in the outer layer of the skin, or epidermis. Far from being an indication of health, a tan is a sign that too much UV exposure has occurred, resulting in damage to the skin.

While the biological function of skin color is fairly straightforward (see Chap. 5 by Lessin and colleagues), the rest of the story certainly is not. Skin color is much more than just a phenotype; it frequently forms the basis for socially constructed definitions of beauty, health, and social status. The allure of tanned skin is unique to Western cultures, and particularly Caucasians. Accordingly, the focus of this chapter will be on tanning in whites and white cultures. However, the practice of cosmetically altering the appearance of one's skin to meet cultural standards of attractiveness is, sadly, endemic in modern societies around the world. For evidence of this, we need look no further than the multibillion dollar industry that has grown up around cosmetic skin-bleaching in many Asian nations [18].

2.3 Beyond the Pale

The cosmetic appeal of the tan is a relatively recent development. For most of history, pale skin has reigned as the archetype of aesthetic beauty in light-complexioned populations [19]. Examples drawn from a variety of literary sources, proverbs, and scriptures provide evidence of many cultures placing an idealized value on white-skinned beauty. Japanese culture has considered white skin an essential characteristic of feminine beauty since antiquity. An ancient Japanese proverb says that "white skin makes up for seven defects"; that is, a woman's light skin is believed to mask undesirable physical features or "uglinesses" [14, 20]. The famous Chinese poetess, Li Ching Chao, wrote about skin so white that it shines through her red veil. In the Bible, King David's fair complexion is emphasized [19]. Renaissance art, which sought to create the most beautiful representation of the human form, portrayed its human figures with light skin and a rosy hue. References to fair-skinned beauties appear in classical works of literature by Homer, Dante, and Goethe. European court poets waxed reverently about necks that were "white as alabaster" or bosoms as "white as snow" [21]. In the early nineteenth century Brothers Grimm fairy tale "Snow White," the beautiful princess is described as having "skin white as snow, lips red as blood, and hair black as ebony" [22]. So great is Snow White's beauty that it incites the jealous and murderous wrath of her wicked stepmother.

Historically, pallor has also served as a physical representation of social class [23]. White face powders have been used for centuries by the upper classes in Europe and Asia to create the appearance of preternaturally white skin. Parasols have long been used by Chinese emperors, European courtiers, and Spanish royalty to maintain pale complexions, and over time have become a universal symbol of privilege and high social status [21]. A tanned appearance, on the other hand, has traditionally been stigmatizing. Derogative references to suntan even appear in the Bible, including this verse from the Song of Solomon 1–6: *Do not stare at me because I am dark, because I am darkened by the sun. My mother's sons were angry with me and made me take care of the vineyards; my own vineyard I have neglected* [24].

The preoccupation with fair skin still prevails in many Asian and Hispanic societies, and within some ethnic minority populations in the U.S. The social significance of light skin can be seen in the language used to describe beauty in these cultures. *Bihaku*, which literally translates to "white beauty," is the term used to describe the porcelain pale complexion that sets the standard of beauty in Japan [18]. In Hindi, the word *gori*, which translates to "fair skinned," is synonymous with the word for beauty [20]. In Thailand, the term *tua dam*, or black body, is commonly used to insult a person of lower social status [25]. Within certain ethnic groups, nuanced skin color hierarchies are commonly established using words that refer to graduated shades of color, a phenomenon anthropologists call "colorism" [26]. In the colloquial slang of Harlem circa 1930, gradations in African-Americans' skin tone were parsed into "high yaller [yellow], yaller, high brown, vaseline brown, seal brown, low brown, dark brown" [27]. Similarly, in India, descriptive terms for variations in skin pigmentation include "fair," "wheatish," and "dusky." As shade lightens on the gradient, social capital increases within these cultures. Perched atop the gradient is "white," which has connotations of purity, cleanliness, and flawlessness. Over the past two decades, the relentless pursuit of a fair complexion has fueled the rapid growth of the skin lightening industry in many Asian markets. Skin-whitening products with names like "White Radiance," "White Perfect," and "Fair and Lovely" are aggressively marketed to men and women alike throughout Asia and the Pacific Islands [20, 28]. Black-market bleaching preparations frequently contain powerful and potentially dangerous bleaching agents (e.g., chromium, mercury, and arsenic), and can produce disfiguring results [25].

2.4 Before the Dawn

Attitudes around sun exposure have varied throughout history. Sun worship was practiced in many ancient civilizations, including the Aztecs, the Inca, the Egyptians, the Romans, and the Greeks. Sun gods were believed to be a source of life [23]. There is also evidence that both the ancient Greeks and Romans employed sunlight in the treatment of diseases. More than 2,000 years ago, the Greeks built a temple to Aesculapius, the god of medicine, on a mountainside facing the sun.

The temple was used for sunbathing and restoring of health [29]. Ancient Greek scientists, Hippocrates and Pythagoras, wrote about the healing powers of the sun. The Romans, too, engaged in sunbathing. Pliny wrote about the daily walks in the sun taken by his elderly friend Spurinna [29]. However, at other times and places in history, the sun was viewed as something to be avoided, even feared.

Before the 1900s, attitudes towards sun exposure in the U.S. bore little resemblance to the culture that has grown up around tanning today. The now-familiar concept of the tan as an index of health and beauty had not yet emerged. Instead, to the extent possible, people took great care to isolate themselves from the sun, through the use of protective clothing, gloves, and wide-brimmed hats [30]. Heavy drapes were standard in Victorian-era homes, providing a physical barrier to natural light. The parasol became a popular fashion accessory for wealthy women in the U.S. and Europe in the later part of the nineteenth century, a trend that is reflected in the art of that era – women with parasols appear regularly on the canvases of Monet, Renoir, and Manet [30].

Since the luxury of sun protection belonged primarily to the privileged upper classes, a pale complexion was a status symbol. Only unfortunate laborers and members of the working class, who had no choice but to "toil" outdoors, had a tanned appearance. For genteel women and men, any sign of sun exposure (i.e., freckles, tan) was considered a regrettable affliction that required corrective action. It was common to see advertisements in popular magazines for bleaching products or at-home preparations that promised to restore the skin to a white condition [29, 30]. A Pond's advertisement from 1912 warned women to "Beware the Kiss of the Sun…The summer girl has no charm as great as the appeal of a complexion as clear, transparent, and like an apple blossom in its delicate soft coloring" [31]. Of note however, many of these bleaching preparations contained toxic ingredients [32].

In the period around 1900, although the association of UV light with skin cancer had yet to be discovered, a host of other physical and mental disorders were attributed by the medical field to sun exposure. The assertion that sunlight had negative health effects was influenced by the European colonization of tropical regions in which large numbers of light-skinned Westerners moved to unfamiliar tropical settings where they often succumbed to mysterious diseases [33]. In the absence of a well-developed understanding of microbes and pathogens, climatic theories of tropical illness became prominent. Central to these theories was the belief that penetration of fair skin by the intense rays of the tropical sun led to deterioration of health. In 1905, the dangers of sunlight were chronicled in *The Effects of Tropical Light on White Men* [34]. In this volume, Dr. Chas Edward Woodruff explicitly suggested a link between electromagnetic radiation and the harmful effects of the sun. He implicated the sun's rays in the etiology of "tropical neurasthenia," a nebulous diagnosis defined by symptoms such as restlessness, irritability, fatigue, memory loss, insomnia, headaches, diarrhea, ulcers, heart palpitations, alcoholism, insanity, and suicide [34]. Subsequently, Dr. Woodruff commented specifically about the U.S. being too sunny to sustain the health of its fair-skinned inhabitants, declaring "The American girl is a bundle of nerves. She is a victim of too much light" [34]. A few years later, Dr. Percy Brown of Harvard Medical School echoed these concerns in a *New York Times* interview when he suggested that chronic exposure to the sun's rays was at the root of "Americanitis," a condition he described as being characterized by "irritability and extreme activity" [35].

Despite these dire warnings, a paradigm shift was beginning to take place in the U.S. and Europe. In the years to follow, a variety of social forces would combine to drown out the lingering "heliophobic" voices of the past and usher in a new era of "sun worship," one that would alter the American relationship with the tan for much of the next century. It would be several decades before warnings about sun exposure were heard again.

2.5 Into the Light

The industrialization of the American workforce set the stage for a gradual reversal of socially constructed attitudes and beliefs towards tanning. In the period roughly between 1880 and 1920, America underwent massive demographic changes as increasing numbers of people left the countryside and moved into urban environments to work in factories [36]. Urban populations swelled even larger under a steady influx of immigrants from Northern Europe [37]. By 1920, for the first time in history, more Americans lived in cities than outside of them, and a new working class lifestyle was born. Pallor, for so long a status symbol, now came to be associated with those who worked in the factories and inhabited the working class slums. Factory work required long hours spent indoors, away from sunlight, with neither the time nor the means for outdoor leisure pursuits. Moreover, the congested parts of the cities were heavily polluted with coal smoke, and thus individuals living in cities received limited amounts of sunlight. The topic of "sunlight starvation" was popular among physicians and scientists at the time, who concluded that the lack of sunlight threatened the health of urban populations [23, 33, 38]. Common medical wisdom held that "Pale, dull, and perverse children get so as a result of dark, gloomy schoolhouses and tenement rooms" [39]. So-called "diseases of darkness" were endemic in the slums; these included tuberculosis and rickets, as well as diseases of "moral depravity," such as alcoholism, depression, and suicide [23, 33]. Indeed, the pallor of tuberculosis was an increasingly common feature of the working class, who lived together in crowded, often unsanitary conditions. Pale skin ceased to be perceived as the picture of health and social status. Instead, it was the color of the feeble, sunlight-starved, lower class. This new association of pale skin with illness and depravity is reflected in the popular fiction of the period; an appearance of "extraordinary pallor" is a defining physical trait of the vampire in Bram Stoker's *Dracula* [33]. In modern American cinema, unnaturally white skin continues to be used as a visual tool for distinguishing villains, as seen in movies like the *Da Vinci Code* and *The Matrix Reloaded* [40].

Against this backdrop of America's changing demographics, the tan shifted markedly from something to be avoided to a physical representation of upward mobility. America's high society groups regularly summered at popular resorts like

Newport Beach and Atlantic City to escape the heat, grime, and congestion of the cities. A "sun-kissed face and bronzed body" symbolized a lifestyle of affluence, recreation, and vacations to the seashore. Recreational swimming gained popularity, and as it did, bathing suits became more streamlined and functional – and more revealing, to the detriment of sun protection [30]. Men and women alike took up new hobbies like golf, tennis, and bicycle riding [41]. New clothing styles emerged to support this life of leisure. Active wear and sporty styles that promoted freedom of movement and exposed more skin replaced the modest, cumbersome garments of the previous century [29].

2.6 No Tan, No Cure

Along a similar timeframe to these changes in American society, medical beliefs regarding UV exposure were also changing due to a number of scientific and medical advancements. The rise of diseases common in urban settings such as tuberculosis and rickets focused an increasing amount of attention on issues of health and hygiene, including the role of sunlight deprivation as a contributing factor. Simultaneously, scientists and physicians in Europe were beginning to understand more about the physical properties of sunlight and its possible role in the treatment of disease.

German scientist Johann Ritter laid the groundwork for the "sun cure" a century earlier, when he discovered UV light in 1801 [42]. Using silver chloride, a colorless crystal which turns black when exposed to light, Ritter demonstrated that solar rays exist beyond the spectrum of visible light, and are capable of producing photochemical effects [43]. At the time, the physiological effects of UV radiation on the human body were not well understood. It would be more than 100 years before the link between UV radiation and skin cancer was made, and even longer before scientists would describe the action spectrum (i.e., wavelengths) of UV radiation capable of inducing sunburn [43]. Another major discovery was made in 1877, by English chemists Thomas Blunt and Arthur Downes, who first identified the bactericidal and fungicidal properties of UV light [30]. Importantly, they demonstrated that the germicidal properties of sunlight were not related to heat, as previously believed, but rather to wavelength, with the shortest wavelengths (i.e., UV) having the greatest effect. In 1882, German scientist Robert Koch discovered that tuberculosis is caused by a bacterium (i.e., tubercle bacilli) and subsequently demonstrated that exposure to direct sunlight killed the bacterium [44].

Perhaps the greatest advance in light therapy was made by Danish physician Niels Finsen, who in 1897 pioneered the technology for the creation of an artificial sunlamp to treat tuberculosis, making it possible for the first time in history to produce "sunlight" on demand (see Fig. 2.1) [23]. Finsen's work is all the more remarkable given that artificial light itself was still a new field. Only a short time before, in 1879, Thomas Edison had invented the incandescent light bulb; and x-rays were first described by Röntgen in 1895 [43]. Finsen's device consisted of a powerful

Fig. 2.1 Circa 1897: Nurses administer artificial light therapy with Finsen's carbon arc lamp at the Finsen Medical Light Institute. Printed with permission from the University of Copenhagen Medical Museum

carbon arc that produced UV radiation, a convex quartz lens to concentrate the radiation on diseased areas of skin, and a water chamber for cooling [45]. Finsen demonstrated that treatment with his Finsen lamp could cure the cutaneous form of tuberculosis, and was awarded the Nobel Prize for this work in 1903 [46]. The Finsen lamp was a predecessor to the first commercially available sunlamps, which would go on to populate homes and businesses all over America. However, Finsen did not live to witness the ensuing commercial popularity of sunlamps – he died in 1904, the year after he received the Nobel Prize.

The pioneering work of Finsen, Koch, Blunt, and Downes ushered in a new era of medical therapy, involving exposure to UV light, either by natural (heliotherapy) or artificial (phototherapy) means. In the pre-antibiotic era, the sanitarium became the preferred treatment for tuberculosis [47]. These were hospitals located in favorable climates where doctors could exploit the bactericidal properties of sunlight by exposing patients with tuberculosis to a daily regimen of rest, fresh air, and maximal sun exposure [48]. The first sanitarium in the U.S. was founded in 1885 by American physician Edward Trudeau in Saranac Lake, New York [49]. It was situated in the Adirondack mountains, thought to be an ideal climate for maximal exposure to fresh air and sunshine. Patients of the Adirondack Cottage Sanitarium reclined on its porches in specially designed lounge chairs, now famously known as Adirondack

Chairs [49]. By the 1920s, sanitaria in the U.S. were regularly applying heliotherapy and phototherapy to both children and adults [29]. A 1926 report in *The American Journal of Public Health* described the treatment of children recovering from tuberculosis at a New York State sanitarium: "In summer the children play and frolic over the extensive hayfields and woodlands. In winter, with only the protection of a loincloth, they ski, coast, and toboggan on the snow covered hills – their delicate, sick bodies in the meantime being rebuilt and hardened by exposure to the direct sunlight." [50].

Meanwhile, in the developing field of pediatrics, there was speculation about the etiology of rickets, a disfiguring bone disease commonly seen in urban-dwelling children. Among other theories, lack of sunlight and fresh air were implicated as potential causes of the disease [51]. Indeed, treatment of rickets with natural "sunbaths" produced good results [52, 53]. Exposure to artificially produced UV radiation from a mercury arc vapor lamp was also shown to be effective for both preventing and treating rickets [47, 54]. Although the biological mechanism was not fully understood until later, the successful use of heliotherapy and phototherapy in the treatment of rickets was celebrated by many in the medical field as conclusive evidence of the healing properties of UV light.

Wartime also contributed to the popularization of the "sun cure." During World War I, it was discovered that sunlight was beneficial for treatment of compound fractures and ulcerated wounds. Heliotherapy was successfully used in military hospitals across Europe to treat the infected wounds of soldiers [55].

2.7 Sunshine of Life

By the 1920s, increasingly broad claims, many of them wildly exaggerated, about the medical benefits of UV light were being reported to the public and within the medical field. Sun exposure was being touted as a panacea for a wide array of conditions beyond tuberculosis and rickets. The curative powers of UV light were catalogued in medical textbooks at the time; one of these textbooks listed 176 non-dermatologic conditions and 73 dermatologic conditions for which therapeutic benefits of UV light had been documented [47]. The "sun cure" was invoked for diseases as diverse as pneumonia, constipation, hypertension, hypotension, cirrhosis, arteriosclerosis, diabetes, gastric ulcers, and obesity. The antiseptic properties of sunlight were also emphasized frequently. In a 1933 *New York Times* article, the editor of the *Medical Journal and Record* was quoted as saying, "It is beginning to be generally believed that the remedial and even curative properties of sunlight are almost [sic] illimitable" [29]. Such claims about the "sun cure" were popularized in the media and rapidly became part of the cultural lexicon. The peak of this media attention occurred between the years 1925 and 1928 [29].

The nascent field of preventive medicine put forth the idea that sun exposure offered prophylactic, as well as curative, benefits [47]. If sunlight was good for sick people, then it must be good for healthy people too. Among other things, UV light was said to boost resistance to colds and infections, aid metabolism, sharpen mental acuity, increase hemoglobin levels, and improve circulation [47]. Accordingly, increasing sun exposure in children and adults became a public health goal. Popular press instructed readers on the benefits of sunlight: "Sunshine is good medicine…it would be hard to find a simpler one, as cheap and easy to take. The more we learn about our sun the more wonderful it becomes" [56]. Americans were advised to make sure that "the living rooms, the workrooms and studies, and your children's playroom are sunlit…the sleeping rooms are freshened daily by a flood of sunshine through an open window" [39]. Public health recommendations suggested that articles of clothing be loosely woven and porous to allow penetration by the sun's rays.

The enthusiasm around UV exposure and tanning extended to babies and children as well. Daily sunbaths were considered to be essential to the healthy development of children, and parents were urged by both medical authorities and the popular press to expose their babies and children to direct sunlight [23, 29]. Medical journals contained detailed instructions on proper implementation of a sunbath regimen. According to a 1926 article in the *American Journal of Public Health*, "We have found that the best results are obtained by telling mothers that they must get their children sunburnt…The infant should be taken out on bright days for a time between the hours of 11 am and 2 pm and the sun allowed to shine directly on its face… The amount of tanning or sunburn may be taken as a rough estimate of the effectiveness of the treatment" [57]. A few years later, in 1933, the *American Journal of Nursing* advised "Sunlight is of great importance to infants…When the infant is a few weeks old, he should be put out-of-doors in the sunlight, for about a half hour at first, and then gradually increasing the time until he may be kept in the sunlight 2 or 3 h daily" [58]. The popular press was equally enthusiastic about sunbaths for children [59]. In 1923, *Literary Digest* declared, "Nothing is better for babies –or grown people either – than a good coat of tan, we have it on good scientific authority" [60]. Sun suits became a popular children's clothing item around this time, as they allowed for maximum exposure of the skin to direct sunlight [61].

It was during this confluence of societal forces and broad medical support that the view of a tan as a sign of good health, attractiveness, and well-being became wide spread in the U.S. This view would go on to become one of the most enduring beliefs about tanning – and as it turns out, one of the most difficult to dispel.

2.8 The Bronze Age

The emphasis on the health benefits of sunbathing created a new niche market for artificial sources of sunlight. The first generation of commercial indoor tanning devices, called "sunlamps" or "health lamps," were marketed to physicians and consumers for use in the home or office beginning in the 1920s [62]. Modeled on the technology of medical devices such as the Finsen lamp, the typical home sunlamp used either a carbon arc or a mercury vapor light source to produce

Fig. 2.2 Commercial giant GE introduced Americans to the concept of "dual purpose lighting" in the 1930s

UV radiation [47]. The early sunlamp technology was plagued by safety problems, including acute sunburns, blistering, and eye injuries [62]. These injurious effects were related to the properties of the UV radiation emissions. The UV spectrum is divided into three categories based on wavelength: UVA wavelengths are the longest (320–400 nm), followed by UVB (280–320 nm), and UVC (200–280 nm) [2]. UVB and UVC wavelengths are much more likely to produce sunburn than UVA, although both UVA and UVB are now understood to be carcinogenic [63]. Sunlamps emitted high amounts of UVB and UVC radiation (i.e., 30–50% UVB; 20% UVC), enough to produce sunburn within only a few minutes [64, 65]. Consequently, sunlamps were relatively inefficient for purposes of cosmetic tanning. Furthermore, achieving an all-over tan was a difficult proposition, because the amount of surface area exposed to the light source was small. Whole-body irradiation could be achieved, rather awkwardly, by incorporating several lamps into a solarium. However, with the mass marketing of indoor tanning salons still 50 years in the future, whole-body exposure to artificial light was impractical for most ordinary Americans [62].

Sunlamps enjoyed their greatest commercial popularity in the period roughly between 1929 and 1934 [29]. The first sunlamp advertisement appeared in the pages of *Vogue* magazine, and marketing would continue throughout the 1930s and 1940s [47]. Sunlamp advertisements commonly touted the health benefits of tanning, capitalizing on the medical establishment's belief in the benefits of UV light [66]. General Electric introduced America to the concept of "dual-purpose lighting" in the 1930s, suggesting that health-promoting sunlamps could be integrated into every American home as a source of both illumination and well-being. The accompanying advertising campaign extolled consumers to "Bask in the health-protecting ultra-violet rays while you read, play bridge, or bathe. Give children ultraviolet radiation in their playrooms" (see Fig. 2.2) [66]. Sunlamps also sprang up in office buildings during the 1920s and 1930s and were believed to have a positive influence on employee health and productivity. The U.S. House of Representatives installed several sunlamps in the House Office Building in 1929 [67]. Physicians also supported the use of sunlamps, even issuing "prescriptions" for home light baths, as exemplified in this case report of a 52 year-old patient suffering from undefined poor health: "At his first visit I found an anemic, very weak man, not equal to any lengthy walks, and looking fully as old as his age suggested. I recommended, in addition to a very mild water cure, daily light baths with the quartz light. After 3 weeks of daily exposure to the ultraviolet rays this patient could hardly be recognized as the same man. He now appeared to be between 35 and 40 years of age, and presented an aspect of blooming health, with rosy cheeks." [68].

In hindsight, it is apparent that scientific and medical data existed that disproved the over-generalized claims of the "sun cure." However, it was the introduction of antibiotics in the last half of the twentieth century that signaled the end of the heliotherapy movement and miraculous claims about UV light. Despite changing medical opinion, the impression held by the general public of UV light as a health-giving tonic proved to be deeply established and difficult to erase. This appears to have, at least in part, been due to events that were occurring outside of the public health and medical fields.

During the 1920s and 1930s, as broad claims about the benefits of UV exposure were endorsed within the medical community, the fashion and cosmetic industries were also quick to capitalize on the vogue of the tan. French designer and fashion icon Coco Chanel is frequently credited with further popularizing the tan in America, after returning from a vacation to the French Riviera with a dark suntan and declaring in the pages of *Vogue* magazine, "The 1929 girl must be tanned…a golden tan is the index of chic" [69]. Indeed, the "tan is beautiful" message was promulgated by numerous magazine advertisements and cosmetics counters in America from the 1920s onward. A historical review of popular women's magazine advertisements from this period found that beginning in 1928, the number of articles and advertisements promoting tanning increased sharply [70]. New tints of makeup were introduced at cosmetics counters to accommodate the trend for darker skin [29]. A *New York Times* article discussed these new formulations in 1929, saying, "There are lotions that call up the walnut stain of gypsy kidnapping tales, there are creams and powders warranted to keep one tanned even under the splash of salt water" [29].

Clothing styles also changed during the 1920s, as sexual mores loosened and young women challenged the traditional roles of womanhood [71]. The new fashion standard allowed for greater skin exposure, making it easier for women to acquire a tan, and to display it. The trend for showing one's tan was described in *New York Times Magazine* in 1929: "Even the fashion plates show the trend, and advertisements have a tendency to feature slender, haughty women of dusky hues, usually clad in white satin dresses, the better to show off their healthy color" [29]. Similarly, in a 1933 *Collier's* article, a debutante elaborated, "It's handsome to be very brown with a light evening gown." [29] Bathing suits got smaller over this time period too, culminating in the introduction of the bikini in 1946 [72].

2.9 Sun Scorched

Not surprisingly, as the tanning craze swept the nation, sunburn emerged as a new public health concern. The Monday morning sunburn was a sign of the times in Manhattan offices, as was described in a *New York Times Magazine* article: "The season is here when New York's office boys and girls take on the color of a season in the Alps" [73]. City health commissioners around the country issued warnings about excessive tanning, and sunburn clinics were set up at area beaches [29]. In the early days, sunbathers who wished to prevent "sun-scorch" did not have much technology at their disposal. A variety of products were purported to be sunburn preventatives, including vegetable oil, cold creams, petroleum jelly, and cellophane blankets; however, none were particularly effective, and in fact, some were counterproductive. Before commercial sunscreens, self-discipline was the only reliable means of preventing sunburn – a careful regimen of progressively longer tanning sessions would eventually produce a fashionable honey-brown glow while reducing the likelihood of sunburn [74]. In 1944, America's first commercial sunscreen, Coppertone Suntan Cream, was introduced, with the tag line "Don't Be a Paleface." A few years later, the sales pitch would be "Tan-don't burn! Get a faster, deeper, tan plus guaranteed sunburn protection" [29].

Even as tanning grew more popular, concerns within the medical field were growing. In 1928 it was discovered that solar radiation causes skin cancer in mice [42, 75]. By the 1930s, UV radiation was widely recognized as a carcinogen in medical circles. The U.S. Public Health Service issued warnings about the risk of sunbathing as early as 1932 [75]. Concerns about serious health risks led to an official reversal of the *Journal of the American Medical Association* editorial position on sun exposure in 1936. The revised position was: "As far as is known, man actually requires only a relatively small amount of sunshine for the maintenance of normal health, and the greatest danger perhaps at the present time lies in too much exposure to sunlight rather than too little" [76]. However, aside from sunburn concerns, prior to World War II there was little recognition of the potential harm of excessive sun exposure in the popular press or in the general public and warnings about the dangers of UV light were rarely publicized [77]. Even if these messages had reached the masses, it is unlikely that they would have altered behavior due to the high level of publicity surrounding the benefits of sun exposure and the lack of an organized health campaign to counter these claims.

2.10 Fry Now, Pay Later

Concern by the medical establishment regarding sun exposure gradually became increasingly organized throughout the 1950s, 1960s, and 1970s, and subsequently increased in public visibility in the 1980s with a number of large-scale sun-protection campaigns being launched such as "National Skin Cancer Prevention and Detection

Week," "The Children's Sun Protection Program," and "Fry Now, Pay Later" [78, 79]. The impetus for these efforts arose from unprecedented increases in the incidence of melanoma, the most lethal form of skin cancer. In the 1930s, melanoma was a rare cancer, with a lifetime risk of 1 in 1,500 people [80]. But by the 1970s, the incidence of melanoma in whites was increasing at an alarming rate of 7% per year [7]. In 1980, lifetime melanoma risk had risen to 1 in 250; by 1985, it was 1 in 150 [80]. Sun exposure was suspected as the primary etiologic factor in the emerging melanoma epidemic, and ominous warnings became increasingly common in both the popular press and medical journals. An analysis of news media coverage of skin cancer topics between 1979 and 2003 revealed a marked increase in news stories about skin cancer between 1979 and the mid-1980s [81]. A content analysis of *New York Times* articles during roughly the same period found a similar trend and noted that melanoma was most often the primary subject of the articles [82]. President Ronald Reagan's treatment for multiple basal cell malignancies pushed skin cancer further into the public eye, creating a spike in newspaper coverage of skin cancer between 1985 and 1988 [81, 82].

In response to the growing public health threat, organizations such as the American Cancer Society, the Skin Cancer Foundation, and the American Academy of Dermatology launched a series of public education campaigns in the mid-1980s around sun safety [79]. Australia had pioneered population-level sun-safety efforts with its highly successful "Slip, Slop, Slap" campaign, which implored citizens to "Slip on a shirt, Slop on sunscreen, and Slap on a hat" when they go outside. In the summer of 1985, the American Cancer Society launched its "Fry Now, Pay Later" campaign, warning Americans about the dangers of sunbathing. Promotional materials showed a deeply bronzed sunbather reclining in a lounge chair, beside a caption that reads: "There is a proven connection between sun exposure and skin cancer, as well as premature wrinkling. If you must be in the sun, use sunscreen and common sense" [78]. In the same year, Ronald Reagan signed a proclamation establishing a National Skin Cancer Prevention and Detection Week. In conjunction with this event, the American Academy of Dermatology began offering free skin cancer screenings through its National Skin Cancer Screening Campaign, in an effort to increase skin cancer awareness and early detection. The program is still in existence and has resulted in the detection of more than 188,000 suspicious lesions, including more than 21,500 suspected melanomas to date [83]. In the years to follow, an increasing number of public education efforts would be undertaken, in the hopes that Americans would change their sunbathing habits [84].

Perhaps as a result of shifting news media coverage and educational campaigns, public awareness of the harmful effects of the sun did increase during this time period, from 48% of the American public in 1985 to 74% in 1996 [12]. However, despite public health efforts to quash the suntan fad during the second half of the twentieth century, the allure of tanned skin remained as strong as ever. Although still relatively few individuals used tanning beds in 1996, the percentage rose from 2% in 1985 to 6% in 1996 and having a tan was still considered to enhance one's appearance [12]. To understand why this is so, it is necessary to examine the cultural context in which these efforts were occurring.

2.11 Dying for a Tan

Even as the medical establishment increasingly distanced itself from purported health benefits of UV exposure and tanning, the social significance of the suntan was becoming more firmly established in America. The 1960s brought with it a new genre of Hollywood film glorifying the beach, the bikini, and the suntan. Beginning with 1959's *Gidget*, the "beach-party movie" genre quickly gained mass appeal with a series of box office hits like *Where The Boys Are*, *Beach Party*, *Bikini Beach*, and *Muscle Beach Party* [74]. The plotlines featured young, bikini-clad, beach bunnies and tan, muscled men enjoying romance and fun in the sun. For the next several decades, Hollywood would turn out a steady stream of golden girls, maintaining the image of the tan as a key element of sensuality and beauty. The original bronzed bombshell, Brigitte Bardot, famously wore a bikini to the Cannes Film Festival in 1953 [72]. In subsequent years, the tanned and toned physiques of numerous celebrities like Jane Fonda, Farrah Fawcett, and Bo Derek would be exemplars of a culture in which the tan was a central feature, inspiring young American women and men to work on their tans.

The growing travel industry also may have contributed to America's tanning practices. Before the 1950s, destination travel was prohibitively expensive except for the wealthiest Americans [23, 74]. The middle class had to settle for excursions within easy reach of home, by railway or steamboat. With the introduction of mass airline travel in the late 1950s, a suntan in the winter became a newly attainable status symbol for many Americans [85]. A 1966 *Newsweek* report suggested that the year-round tan could be "worn like Brooks Brothers clothes as a sure sign of affluence" [85]. Travel by automobile, bus, and recreational vehicle were also more affordable than ever, opening up new horizons for Americans of ordinary means [74]. Families across America embraced the "family vacation," often planning summer holidays or winter getaways long in advance. The suntan was a valued souvenir that one could bring back from vacation (see Fig. 2.3).

Meanwhile, tips about the best way to achieve a tan proliferated. A 1964 magazine article in *Mademoiselle* offered readers detailed instructions on how to achieve the

Fig. 2.3 In the 1950s and 1960s, Americans enjoyed unprecedented access to sunny locales, making it even easier to maintain a year-round tan

"perfect tan" using the "rotisserie" method, which involved assuming different sunbathing positions at regular intervals to ensure maximum sun exposure. The article recommended the use of face and body reflectors to intensify the sun's rays and advised readers to "seek the noonday sun" [86]. Indeed, fanatical sunbathers engaged in all sorts of strategies to optimize tanning, including coating themselves in a mixture of baby oil and iodine, rubbing their skin with salt, and wearing aluminum reflectors around their necks [85]. Also available was a virtual apothecary of commercial suntan products, all promising to produce faster, better results. A 1977 Coppertone advertisement boasted: "Coppertone helps more people get a magnificently deep fast tan than any other suntan product in the whole world. And Coppertone does even more. A special ingredient in Coppertone actually helps keep your skin looking young while you get a great tan" [87].

Popular culture's emphasis on the attractiveness of tans was also reflected in toys. In the 1970s, Mattel, Inc. introduced its sun-friendly "Malibu Barbie" line, featuring dolls like Sunsational Malibu Barbie, Sun Gold Malibu Barbie, and Sun Loving Malibu Barbie with peek-a-boo tan (i.e., tan lines). Around the same time, the Ideal Toy Company introduced the "Suntan Tuesday Taylor" doll, which had the unique ability to "tan" when exposed to the sun. According to an advertisement for the doll: "Put her in the sun for just a minute…her skin turns to a deep golden tan. And just like a real tan, it gradually fades when you keep her out of the sun for an hour. Look at the skin under her bathing suit strap, which hasn't been exposed to the sun or light. The difference is amazing! Trim, tan, and terrific…that's Suntan Tuesday Taylor" [88].

2.12 Selling the Suntan

The latter part of the twentieth century also witnessed the rapid growth of two powerful industries: the sunscreen industry and the indoor tanning industry. On the surface, these two industries would seem to be diametrically opposed. However, a more careful analysis reveals that both were in the business of selling exposure to UV radiation, and thus contributing to the persistence of tanned skin as the cultural standard of attractiveness.

Sunscreen, when used appropriately, can effectively protect skin from the harmful effects of sun exposure, including some skin cancers [89]. Importantly, safe-sun guidelines developed by scientific consensus panels recommend the use of sunscreen with a sun protection factor (SPF) of 15 or greater [79]. In this regard, the sunscreen industry represents a possible positive cultural influence on efforts to increase population-level sun protection in the U.S. Historically, the sun care industry has also played an instrumental role in promoting sun exposure. The first suntan products appeared in the 1940s. Their purpose was to screen out burning rays while allowing the user to develop a tan; however, early products varied greatly in their effectiveness [75]. As formulas improved over the next three decades, suntan products enjoyed increasing popularity as tanning aids. The average consumer was only interested in sun protection to the extent that it could promote a painless, perfect tan. The majority of the sun care market consisted of deep tan lotions or oils, containing minimal sunscreen; cocoa butter and baby oil were also bestsellers as promoters of tans, despite having no sunscreen properties at all. Products that promised to intensify or encourage a deeper, darker tan would dominate the industry until the late 1970s.

In the 1980s, as the medical community's messages about skin cancer increased, the industry responded by modifying its marketing approach. Capitalizing on consumer concerns about skin cancer, as well as a new Food and Drug Administration (FDA)-mandated coding system for sunscreen (i.e., SPF), advertising tactics shifted from products that "tan you" to products that "protect you from sunburn." Sunscreens quickly became the most rapidly growing segment of the sun care industry, increasing by 70% between 1980 and 1985, as consumers jumped at the opportunity to "tan safely" [29]. With established brand names like Bain de Soleil (i.e., sunbath), Hawaiian Tropic, and Coppertone, sunscreen products continued to promote the idealized image of bronze skin, now presented as *healthy* bronze skin. In an analysis of photographs in popular women's magazines between 1983 and 1993, models in sunscreen advertisements had darker tans, more exposed skin, and wore fewer hats, compared to other types of photographs or advertisements [90]. The underlying message was that Americans could have it all – sun exposure, sun protection, and a healthy glow. The message of a "safe tan" has endured and helped to propel the sunscreen industry to its current behemoth proportions. Sunscreen sales in the U.S. grew from $18 million in 1980 to more than $500 million in 2006 [91]; in 2008, the U.S. sun care market, including sunscreens and all other sun care products, generated total revenues of $1.1 billion [92]. Despite the sun protective benefits associated with sunscreen use, studies show that most people who use sunscreen do not apply enough of the product to achieve adequate photoprotection [93–97]. Furthermore, among individuals seeking a tan, there is evidence to suggest that sunscreen may actually increase total UV exposure by alleviating sunbathers' concern about extending the amount of time that they spend in the sun [89, 98].

Following the widespread popularity of sunlamps in the 1920s and 1930s, indoor tanning entered a period of relative dormancy. Although people continued to use sunlamps, their use never again reached the levels seen during the heliotherapy craze [29]. Furthermore, the fact remained that sunlamps were not particularly efficient tanning devices. However, although not yet widely marketed to consumers, a series of technological advances had been occurring in the field of artificial UV light which significantly improved the outlook for the indoor tanning industry. The 1960s witnessed the development of low-pressure fluorescent tubes capable of delivering UV radiation for tanning purposes. This technology represented an improvement over the old mercury vapor and carbon arc sunlamps, but still emitted relatively high amounts of UVB radiation. In the U.S., these were the variety of tanning lamps commonly used in the first-generation of tanning booths and beds [62, 99].

In 1978, the first indoor tanning salon opened its doors in Searcy, Arkansas [29]. "Tantrific" was the brainchild of three entrepreneurs, who seized the opportunity to capitalize on the current fashion for a deep, dark tan. The salon itself was nothing more than three stand-up tanning booths, located inside half of an old converted house [100]. The early stand-up tanning booths were approximately 3 ft^2, lined with fluorescent sunlamps and Mylar or aluminum reflectors [101]. Developing a dark tan required multiple 15-min sessions; but for the price of around $2 per visit, customers were happy to bake themselves to a deep golden brown [102].

From these humble beginnings, the indoor tanning industry quickly gathered momentum. Franchises became available for between $7,000 and $35,000 [103]. The typical franchise arrangement provided the tanning equipment and tropical-themed décor, which often included palm trees, rattan furniture, and thatched-roof huts [100]. The growth of this industry was astounding. By December 1979, Tantrific had opened franchises in 81 locations, and were projecting an additional 3,000 outlets [102]. Within 18 months of opening its doors, Tantrific found itself competing with at least a dozen other large chains, including Plan-A-Tan, Tantalize, and Tan Four Seasons [100]. By the spring of 1980, there were an estimated 1,000 tanning salons operating around the country [29]. That same year, it was reported in the *Journal of the American Medical Association* that tanning salons were opening at a rate of 10 per day [101]. Much of the initial growth in the industry occurred in health- and fitness-related businesses, with existing businesses purchasing one or two tanning units as an add-on, rather than developing free-standing salons [104]. Special introductory pricing and referral bonuses were other strategies used to build business [101].

In the 1980s and 1990s, amid growing concern about potential carcinogenic effects of UVB radiation, new technology was developed that shifted the UV outputs of low-pressure fluorescent lamps to UVA [2]. These high intensity lamps were capable of delivering UVA light up to five times as intense as that of normal sunlight, with minimal UVB emissions; the net effect was increased tanning efficiency and decreased burning [65]. Around 1983, tanning devices using the new UVA technology became available in tanning salons around the country [105]. The tanning industry promoted the new "UVA only" tan as a "safe tan," and credits the technology with stimulating even stronger commercial growth [104]. The new tanning beds were also more comfortable than their predecessors, because they utilized a clamshell design that allowed customers to recline while they tanned [29]. In reality, the "safe tan" claim was not supported by the evidence. Many of the new "UVA only" lamps still emitted appreciable amounts of UVB rays, and thus continued to pose a risk of UVB-induced sunburn [106]. Moreover, "UVA only" lamps carried their own set of risks, related to the high intensity outputs of UVA radiation, which like UVB radiation causes skin cancer and photoaging [107]. Yet, with tanning being marketed as increasingly comfortable, "safe," and efficient, it is no surprise that the industry continued to experience rapid growth during the 1980s and 1990s. In 1986, there were an estimated 18,000 tanning salons around the country; by the beginning of the new millennium, at least 50,000 tanning facilities existed [29, 108].

As with sun-based UV exposure, increased sunbed use was accompanied by increased warnings from the medical establishment about the dangers of sunbeds [29, 109]. Fashion magazines were also beginning to change their emphasis. A historical analysis of photographs in American girls'/women's magazines between 1983 and 1993 showed a trend towards lighter tans, more women wearing hats, more sunscreen advertisements, and sun awareness articles [90]. However, the growing recognition of the consequences of sun exposure in the 1980s and 1990s was juxtaposed with continued glorification of the tan in popular culture. Don Johnson's iconic style in the series *Miami Vice* (1984–1989) was defined by a deep tan, pastel shirts, and white linen pants. In a famous 1991 Pepsi commercial, supermodel Cindy Crawford's golden glow is prominently featured beneath a white tank top and cutoff shorts. The mega-hit television series *Baywatch*, which ran from 1989 to 1999, gained instant fame around the world for its cast of buxom, tanned lifeguards in red swimsuits.

2.13 The Age of Contradiction

With the arrival of the new millennium has come widespread recognition of the dangers of a tan. Several decades of media campaigns have successfully established the link between tanning and skin cancer in the collective conscience of the American public [10]. The media is no longer explicitly emphasizing tans, and in some cases, they are outright rejecting them. Gone are the days of helpful tanning tips and strategies in the popular press; now, magazines routinely publish articles about the dangers of UV exposure and how best to protect oneself from the sun, with explicit recognition of the idea that there is no such thing as a safe tan. In 2006, *Cosmopolitan* magazine even took things one step further, by launching the Practice Safe Sun campaign, a skin cancer prevention initiative that asked readers to stop all forms of tanning, aside from tans "in a bottle" referring to sunless tanning products [110]. However, this in itself highlights the discrepancy between concerns about dangerous UV exposure and the continued cultural value of the tan. Sunless tanning is increasing in popularity and provides a safer alternative to UV tanning, but sunless tans are not a method of sun protection and promote the image of a tan [111, 112]. In addition, according to industry information, 38% of indoor tanning salons in 2008 offered at least one sunless booth [113]. By allowing clients to choose the tanning method depending on their need, clients had the potential for switching back and forth between UV and sunless tans. The impact of this as a means to promote continued UV exposure is only now beginning to be explored within the public health research community (see Chap. 11 by Pagoto).

Sufficient epidemiological evidence of cancer risk has now accumulated to result in the classification of UV-emitting tanning devices as carcinogenic to humans by the International Agency for Research on Cancer [114]. Nonetheless, the demand for the quickie tan remains high. According to industry estimates, 30 million Americans visit tanning salons each year [115]. A recent national survey found that 10% of children between the ages of 11–18 had used an indoor tanning facility during the last year, with much higher use found in subgroups of teens (e.g., girls ages 14–17) [116]. Much like the tobacco industry, which also profits from selling a carcinogenic product to consumers, the indoor tanning industry has developed a set of sophisticated techniques to counter public health warnings and fuel consumer demand for the perfect tan (see Chap. 10 by Suarez and colleagues). Today, the indoor tanning industry earns $5 billion in estimated annual revenue, which represents a large increase from $1 billion in 1992 [106, 113]. There are 25,000 free-standing indoor tanning salons nationwide, and an additional 20,000 outlets in business establishments such as health clubs, spas, etc. [115]. In many of the largest U.S. cities, tanning salons outnumber McDonalds and Starbucks [117]. In addition to indoor tanning salons, there are around a dozen manufacturers of sunlamps/sunbeds who sell their products in the U.S. to the tune of $72 million per year. Interestingly, more than 50% of tanning salon business owners are women, as opposed to the 25% ownership rates typically seen in other industries [118].

A double standard appears to have developed around tanning in the popular media and press. A content analysis of articles in eight popular women's magazines conducted between 1997 and 2006 found that coverage of tanning risks frequently occurs alongside mentions of tanning benefits [119]. Across all magazines, benefits of tanning received regular coverage, with "looking healthy" as the benefit mentioned most often; other purported benefits of tanning included "looking attractive," "looking sexy," "looking thin," and "having fun." Interestingly, the same magazine that is now leading the charge on sun safety, *Cosmopolitan*, featured benefits of tanning more often than any of the other magazines. Linda Wells, the editor of fashion magazine *Allure* declared in a 2001 *New York Times* article, "The deep, dark tan isn't as attractive to the upper classes. It looks cheap; it demonstrates a lack of control" [120].

Indeed, in the twenty-first century, the overly dark or otherwise fake-looking suntan (i.e., orange) has taken on stigmatizing connotations. The natural tan is now the fashion; that is, a healthy glow that happens by accident, or at least appears that it could have. Thirty years of health warnings about the dangers of UV exposure seems to finally be shifting social perceptions of intentional tanning, from a healthy behavior to an unhealthy one. Like the stigma that has come to be associated with smoking, intentional sun exposure leading to a deep tan is increasingly viewed as irresponsible, ignorant, or both. Slang terms such as "fake-baking" (i.e., using an indoor tanning salon) "perma-tan" (i.e., an unnatural looking year-round tan) and "tanorexia" (i.e., a "disease" in which no matter how tan a person is, it is never tan enough) have entered the cultural lexicon, and are commonly used to shame overzealous tanners who have taken the tanning habit too far [121–123]. Artificial-looking tans, whether obtained from UV exposure or a bottle (see Chap. 11 by Pagoto), are often viewed a sign of weakness in which a person succumbs to vanity. During the recent presidential campaign, the revelation that vice presidential candidate Sarah Palin had installed a tanning bed in the Alaska governor's mansion provided endless fodder for jokes about her perceived vanity. Hollywood icon George Hamilton, whose preternaturally tanned skin is emblematic of the "perma-tan," is regularly lampooned in the media, so much so that his orange glow has become a running joke in popular culture. In 2009, *The Boston Globe* ran an article entitled, "You're not bronzed! That tan is fake!" detailing the increasingly common fashion faux pas of an unnaturally orange glow. The article gently scolds, "They may think they resemble a deeply bronzed Cindy Crawford on a Palm Springs golf getaway, but the truth is, these chronic tanaholics look artificial…they wind up as tangerine-colored as Donald Trump" [124]. Unfortunately, embedded in the backlash against the artificial tan is the unmistakable message that a tan is desirable, as long as it is natural-looking.

2.14 Light on the Horizon

Media coverage and public health campaigns have been successful in raising awareness of the health risks of UV exposure and indoor tanning [10, 12, 125]. However, these efforts have not led to significant improvements in skin protection behaviors [12, 126, 127]. Recent research suggests that interventions that attend to individuals' concern about their appearance may be more effective in countering the pervasive normative influences on tanning behavior than educational campaigns alone [127]. In this regard, appearance-based interventions seem to represent a promising avenue for changing behavior. Women and girls are a particularly important target for such interventions, given the societal value that is placed on appearance, weight, and youth among these populations [128–131]. The indoor tanning industry has aggressively marketed an unrealizable standard of beauty to adolescent girls and young women, exploiting females' body image insecurities and desire to be sexually appealing [131]. Tanning advertisements draw associations between a tan body and an attractive body, featuring beautiful thin models with slogans like "Welcome To A Better Looking You" [132]. Popular colloquialisms like, "tan fat looks better than white fat," and, "if you can't tone it, tan it," confirm that this message is being heard loud and clear by American girls and women.

As has been clearly demonstrated within the field of tobacco control, a crucial factor in addressing any significant public health issue is the institution of effective public policy. Regulatory and legal approaches restricting access to tanning facilities, when appropriately enforced, have been shown to be effective in reducing UV exposure (see Chap. 10 by Suarez and colleagues) [133, 134]. Some U.S. states have passed legislation mandating regulation of indoor tanning facilities in an effort to protect consumers [133]. However, inconsistencies in legal restrictions and enforcement across states in addition to a number of states failing to gain regulatory authority over tanning facilities has led to the call for national legislation of indoor tanning to protect minors [135].

More stringent federal regulation of indoor tanning may be on the horizon. The Tanning Bed Cancer Control Act was introduced into the U.S. House of Representatives by Rep. Carolyn Maloney, (D-NY) and Rep. Charlie Dent (R-PA) in January 2010; if passed, the law would expand federal regulation of tanning limiting the amount of UV rays emitted by tanning beds and the length of time a consumer can be exposed to the harmful radiation. It would also set a minimum age limit of 18 for tanning bed users. Meanwhile, the FDA's Medical Devices Advisory Committee is in the process of reevaluating the classification of UV-emitting tanning devices [136]. The FDA could potentially reclassify tanning beds as a more serious type of medical device that would require greater regulatory controls.

Currently, tanning beds are classified as Class I medical devices, the same classification as elastic bandages. Finally, beginning on July 1, 2010, the U.S. government began imposing a 10% excise tax on indoor tanning services. The "tanning tax" is a provision of the sweeping health care overhaul legislation that was signed into law in March 2010. It remains to be seen whether these policy efforts will succeed in diminishing America's relentless pursuit of the perfect tan.

2.15 Conclusion

This chapter has reviewed the key events related to the current context in which indoor tanning occurs. During the twentieth century, Americans' exposure to UV radiation has increased dramatically. A number of factors have contributed to this change, including the industrialization of America, changing opinions and practice within medical science, the advent of inexpensive mass travel, the establishment of the tan as a core fashion element, and the influences of various industries that profited from selling tans. In particular, the indoor tanning industry has played an essential role in promoting the tan, beginning with the introduction of the first sunlamps in 1920. Its subsequent rapid growth in the late 1970s and 1980s and expanded marketing of artificial UV exposure are considered by some to be a causal factor for increasing lifetime UV exposure in generations of Americans [65]. The high-intensity, repeated UV exposures that are accumulated through indoor tanning represent a significant departure from the baseline levels of exposure experienced over most of human history [63]. For this reason, it is still unclear what the long-term consequences of indoor tanning will be on population health. Sadly, it is already becoming clear that this increased UV exposure has been accompanied by concomitant increases in skin cancer incidence, particularly among young women [9, 137]. The social constructions of the past, which equate bronzed skin with health and attractiveness, have elevated the tan to a position of undeserved status. Current attitudes and behaviors around tanning need to change, if we hope to successfully combat the epidemic of skin cancer. Without intervention, these cultural norms will be perpetuated in yet another generation of tan-seekers.

References

1. English DR, Armstrong BK, Kricker A et al (1997) Sunlight and cancer. Cancer Causes Control 8(3):271–283
2. IARC. Solar and ultraviolet radiation. International Agency for Research on Cancer 2007
3. Tucker MA (2009) Melanoma epidemiology. Hematol Oncol Clin North Am 23(3):383–395
4. Altekruse SF, Kosary CL, Krapcho M et al (2010) SEER cancer statistics review, 1975–2007. National Cancer Institute, Bethesda
5. Christenson LJ, Borrowman TA, Vachon CM et al (2005) Incidence of basal cell and squamous cell carcinomas in a population younger than 40 years. JAMA 294(6):681–690

6. Gandini S, Sera F, Cattaruzza MS et al (2005) Meta-analysis of risk factors for cutaneous melanoma: II. Sun exposure. Eur J Cancer 41(1):45–60
7. Jemal A, Devesa SS, Hartge P et al (2001) Recent trends in cutaneous melanoma incidence among whites in the United States. J Natl Cancer Inst 93(9):678–683
8. Linos E, Swetter SM, Cockburn MG et al (2009) Increasing burden of melanoma in the United States. J Invest Dermatol 129(7):1666–1674
9. Purdue MP, Freeman LE, Anderson WF et al (2008) Recent trends in incidence of cutaneous melanoma among US Caucasian young adults. J Invest Dermatol 128:2905–2908
10. Hay J, Coups EJ, Ford J et al (2009) Exposure to mass media health information, skin cancer beliefs, and sun protection behaviors in a United States probability sample. J Am Acad Dermatol 61(5):783–792
11. Keeney S, McKenna H, Fleming P et al (2009) Attitudes, knowledge and behaviours with regard to skin cancer: a literature review. Eur J Oncol Nurs 13(1):29–35
12. Robinson JK, Rigel DS, Amonette RA (1997) Trends in sun exposure knowledge, attitudes, and behaviors: 1986 to 1996. J Am Acad Dermatol 37(2 Pt 1):179–186
13. Weinstein JM, Yarnold PR, Hornung RL (2001) Parental knowledge and practice of primary skin cancer prevention: gaps and solutions. Pediatr Dermatol 18(6):473–477
14. Holubar K, Schmidt C (1998) Historical, anthropological, and biological aspects of sun and the skin. Clin Dermatol 16(1):19–22
15. Jablonski NG, Chaplin G (2000) The evolution of human skin coloration. J Hum Evol 39(1): 57–106
16. Harris M (1998) Our kind: who we are, where we came from, and where we are going. HarperCollins Publishers, Inc., New York
17. Gange RW (1986) Tanning. Dermatol Clin 4(2):189–193
18. Malkan S (2007) Not just a pretty face. New Society Publishers, Gabriola Island
19. Holubar K (1997) Black skin and the ideal of beauty. J Invest Dermatol 108(5):813–814
20. Wagatsumi H (1967) Social perception of skin color in Japan. Daedalus 96(2):407–443
21. Harris M (1973) The rites of summer. *Nat Hist* 82:20
22. Brothers Grimm (2010) Fairy tales, history, facts and more. http://www.nationalgeographic.com/grimm/.National Geographic. Accessed 5 Apr 2010
23. Randle HW (1997) Suntanning: differences in perceptions throughout history. Mayo Clin Proc 72(5):461–466
24. Holubar K (1998) Photodamage: an historical perspective. J Investig Dermatol Symp Proc 3(1):45–46
25. Asian Pacific Post (2006) Asia's obsession with white skin leaves a black mark. http://www.asianpacificpost.com/portal2/ff8080810b1faf95010b2498f44a01b7_Asian_white_skin.do.html. Accessed 28 Oct 2010
26. Thompson MS, Keith VM (2004) Copper brown and blue black: colorism and self evaluation. In: Herring C, Keith V, Horton HD (eds) Skin deep: how race and complexion matter in the "Color Blind" era. University of Illinois Press, Champaign
27. Hurston ZN (2010) Glossary of Harlem slang: slang terms circa 1930s. http://aalbc.com/authors/harlemslang.htm. Accessed 3 Mar 2010
28. Li EPH, Min HJ, Belk RW et al (2008) Skin lightening and beauty in four Asian cultures. Adv Consum Res 38:444–449
29. Segrave K (2005) Suntanning in 20th century America. McFarland & Company, Inc., Jefferson
30. Albert MR, Ostheimer KG (2002) The evolution of current medical and popular attitudes toward ultraviolet light exposure: part 1. J Am Acad Dermatol 47(6):930–937
31. Pond's Vanishing Cream (Pond's Extract Company advertisement). John W. Hartman Center for Sales, Advertising and Marketing History 1912
32. Koblenzer CS (1998) The psychology of sun-exposure and tanning. Clin Dermatol 16(4): 421–428
33. Carter S (2007) Rise and shine: sunlight, technology, and health. Berg, Oxford

34. Woodruff C (1905) The effects of tropical light on white men. Rebman Company, New York
35. Our energy due to the sun. *The New York Times* 1911
36. Hirschman C, Mogford E (2009) Immigration and the American industrial revolution from 1880 to 1920. Soc Sci Res 38(4):897–920
37. Gibson C, Jung K (2006) Historical census statistics on the foreign-born population of the United States: 1850–2000. Working Paper No. 81. http://www.census.gov/population/www/documentation/twps0081/twps0081.pdf. Accessed 9 Aug 2010
38. Stradling D (1999) Smokestacks and progressives: environmentalists, engineers, and air quality in America, 1881–1951. Johns Hopkins University Press, Baltimore
39. McGovern (1936) The secret of keeping fit: an easy and sure way to better health. Simon and Schuster, New York
40. Chan C, Wagner RF Jr (2009) Dermatology at the movies. Clin Dermatol 27(4):419–421
41. Bolanca Z, Bolanca I, Buljan M et al (2008) Trends, habits and attitudes towards suntanning. Coll Antropol 32 (Suppl 2):143–146
42. Urbach F, Forbes PD, Davies RE et al (1976) Cutaneous photobiology: past, present and future. J Invest Dermatol 67(1):209–224
43. Hockberger PE (2002) A history of ultraviolet photobiology for humans, animals and microorganisms. Photochem Photobiol 76(6):561–579
44. Blevins SM, Bronze MS (2010) Robert Koch and the 'golden age' of bacteriology. Int J Infect Dis 14(9):e744–e751
45. Lawrence G (2002) Tools of the trade: the Finsen Light. Lancet 359(9319):1784
46. Mayer E (1932) The curative value of light. Appleton, New York
47. Albert MR, Ostheimer KG (2003) The evolution of current medical and popular attitudes toward ultraviolet light exposure: part 2. J Am Acad Dermatol 48(6):909–918
48. Potts J (1931) The effect of sunlight on man. Phi Delta Kappan 14(1):29–32
49. Lindberg DA, Howe SE (2009) My flying machine was out of order. Trans Am Clin Climatol Assoc 120:99–111
50. Fitzgerald RE (1926) Perrysburg – a mecca for heliotherapy. Am J Public Health (NY) 16(9):893–895
51. Rajakumar K (2003) Vitamin D, cod-liver oil, sunlight, and rickets: a historical perspective. Pediatrics 112(2):e132–135
52. Hess AF, Unger LJ (1921) The cure of infantile rickets by sunlight: preliminary note. JAMA 77:39
53. Rajakumar K, Greenspan SL, Thomas SB et al (2007) SOLAR ultraviolet radiation and vitamin D: a historical perspective. Am J Public Health 97(10):1746–1754
54. Holick MF (2008) Sunlight, UV-radiation, vitamin D and skin cancer: how much sunlight do we need? Adv Exp Med Biol 624:1–15
55. Lim H, Honigsman H, Hawk J (2007) Photodermatology. Informa Healthcare, New York
56. Sunshine as medicine. *Literary Digest* 1924; 97:53
57. Tisdale F (1926) Sunlight and health. Am J Public Health 16(7):697–698
58. Mayer E (1933) Sunbathing and sunlamps. Am J Nurs 33(8):739–746
59. Babies and sunshine. *Literary Digest* 1923; 79:27
60. Old Doctor Sun. *Literary Digest* 1925; 87:28
61. Sunshine and clothing (editorial) JAMA 1929; 106:2071–2072
62. Diffey BL (1995) Cosmetic tanning and human skin cancer. In: Mukhtar H (ed) Skin cancer: mechanisms and human relevance. CRC Press, Boca Raton
63. Ibrahim SF, Brown MD (2008) Tanning and cutaneous malignancy. Dermatol Surg 34(4): 460–474
64. Diffey BL, Farr PM, Ferguson J et al (1990) Tanning with ultraviolet A sunbeds. BMJ 301(6755): 773–774
65. Cesarini J (2007) Solaria. In: Ringborg U, Brandberg Y, Breitbart E, Greinert R (eds) Skin cancer prevention. Informa Healthcare, New York
66. Albert MR (2000) Dermatology related advertisements from the 1920s and 1930s. Int J Dermatol 39(5):389–395

67. Sun lamps to tan cheeks of members of Congress. The New York Times 1929. p 1
68. Lorand A (1925) Life shortening habits and rejuvenation. FA Davis Company, Philadelphia
69. Chapman S, Marks R, King M (1992) Trends in tans and skin protection in Australian fashion magazines, 1982 through 1991. Am J Public Health 82(12):1677–1680
70. Martin JM, Ghaferi JM, Cummins DL et al (2009) Changes in skin tanning attitudes. Fashion articles and advertisements in the early 20th century. Am J Public Health 99(12):2140–2146
71. Kitch C (2001) The girl on the magazine cover: the origins of visual stereotypes in American mass media. University of North Carolina Press, Chapel Hill
72. Turner J (2006) A brief history of the bikini. Slate Magazine. http://www.slate.com/id/2145070. Accessed 5 Apr 2010
73. Now we go weekend-ing. *The New York Times Magazine* July 1, 1928. p 72
74. Lencek L, Bosker G (1999) The beach: history of paradise on earth. Penguin, New York
75. Albert MR, Ostheimer KG (2003) The evolution of current medical and popular attitudes toward ultraviolet light exposure: part 3. J Am Acad Dermatol 49(6):1096–1106
76. Sunlight and health. *Can Med Assoc J* 1936; 35(5):571–572
77. Keesling B, Friedman HS (1987) Psychosocial factors in sunbathing and sunscreen use. Health Psychol 6(5):477–493
78. Grossman DJ (1989) Public and professional educational materials on skin cancer. J Am Acad Dermatol 21(5 Pt 1):1012–1018
79. Koh HK, Geller AC (1998) Public health interventions for melanoma. Prevention, early detection, and education. Hematol Oncol Clin North Am 12(4):903–928
80. Rigel DS, Carucci JA (2000) Malignant melanoma: prevention, early detection, and treatment in the 21st century. CA Cancer J Clin 50(4):215–236, quiz 237–240
81. Stryker JE, Solky BA, Emmons KM (2005) A content analysis of news coverage of skin cancer prevention and detection, 1979 to 2003. Arch Dermatol 141(4):491–496
82. Heneghan MK, Hazan C, Halpern AC et al (2007) Skin cancer coverage in a national newspaper: a teachable moment. J Cancer Educ 22(2):99–104
83. American Academy of Dermatology (2010) History of the skin cancer screening program. http://www.aad.org/public/exams/screenings/. Accessed 5 Apr 2010
84. Edlich RF, Winters KL, Cox MJ et al (2004) National health strategies to reduce sun exposure in Australia and the United States. J Long-Term Eff Med Implants 14(3):215–224
85. Pendergast T, Pendergast S (2000) St. James encyclopedia of pop culture. St. James Press, Detroit
86. Mlle plots the perfect tan. *Mademoiselle* 1964
87. Beautiful tan today, young looking skin tomorrow (inside front cover). Ms Magazine 1972
88. Ideal toy buyers catalog. *Suntan Tuesday Taylor* 1977; 1261–1267
89. Autier P (2009) Sunscreen abuse for intentional sun exposure. Br J Dermatol 161(Suppl 3): 40–45
90. George PM, Kuskowski M, Schmidt C (1996) Trends in photoprotection in American fashion magazines 1983–1993. Will fashion make you look old and ugly? J Am Acad Dermatol 34(3): 424–428
91. Lawyers USA (2006) Lawsuit filed against sunscreen manufacturers for deceptive advertising. http://www.allbusiness.com/services/legal-services/4086005-1.html. Accessed 5 Apr 2010
92. Datamonitor (2010) Personal hygiene/personal care: United States industry guide. http://www.researchandmarkets.com/reportinfo.asp?report_id=889057. Accessed 5 Apr 2010
93. Bech-Thomsen N, Wulf HC (1992) Sunbathers' application of sunscreen is probably inadequate to obtain the sun protection factor assigned to the preparation. Photodermatol Photoimmunol Photomed 9(6):242–244
94. Diffey BL (1996) Sunscreens, suntans and skin cancer. People do not apply enough sunscreen for protection. BMJ 313(7062):942
95. Neale R, Williams G, Green A (2002) Application patterns among participants randomized to daily sunscreen use in a skin cancer prevention trial. Arch Dermatol 138(10):1319–1325
96. Stenberg C, Larko O (1985) Sunscreen application and its importance for the sun protection factor. Arch Dermatol 121(11):1400–1402

97. Wright MW, Wright ST, Wagner RF (2001) Mechanisms of sunscreen failure. J Am Acad Dermatol 44(5):781–784
98. Naylor M, Robinson JK (2005) Sunscreen, sun protection, and our many failures. Arch Dermatol 141(8):1025–1027
99. Spencer JM, Amonette RA (1995) Indoor tanning: risks, benefits, and future trends. J Am Acad Dermatol 33(2 Pt 1):288–298
100. Sun salons. *Time* 1980. p 59
101. Kaminester LH (1980) Suntanning centers. JAMA 244(11):1258–1259
102. 90 second suntans. *Newsweek* 1979. p 68
103. Tanning parlors fade from franchise scene. *The New York Times* May 26, 1981
104. Indoor Tanning Market. http://www.lookingfit.com/articles/indoor-tanning-market-factbook-2009.html. Accessed 5 Apr 2010
105. US Food and Drug Administration (2010) FDA guidance document: inspection and field testing of radiation-emitting electronic products – Attachment C: specific instructions for sunlamp product inspections and tests. http://www.fda.gov/MedicalDevices/DeviceRegulationandGuidance/GuidanceDocuments/ucm112671.html. Accessed 30 July 2010
106. Rivers JK, Norris PG, Murphy GM et al (1989) UVA sunbeds: tanning, photoprotection, acute adverse effects and immunological changes. Br J Dermatol 120(6):767–777
107. Hawk J (2000) Sunbeds. Radiat Prot Dosimetry 91(1–3):143–145
108. Kwon HT, Mayer JA, Walker KK et al (2002) Promotion of frequent tanning sessions by indoor tanning facilities: two studies. J Am Acad Dermatol 46(5):700–705
109. American Cancer Society (1981) Are tanning centers safe? *Cancer News* 19 (Spring/Summer)
110. Smith SD (2006) Cosmo to promote safe skin. *Mediaweek*. http://www.allbusiness.com/services/business-services-miscellaneous-business/4775142-1.html. Accessed 5 Apr 2010
111. Fu JM, Dusza SW, Halpern AC (2004) Sunless tanning. J Am Acad Dermatol 50(5):706–713
112. Stryker JE, Yaroch AL, Moser RP et al (2007) Prevalence of sunless tanning product use and related behaviors among adults in the United States: results from a national survey. J Am Acad Dermatol 56(3):387–390
113. Looking Fit (2010) Facts and figures. http://www.lookingfit.com/articles/08factp13.pdf. Accessed 30 July 2010
114. El Ghissassi F, Baan R, Straif K et al (2009) A review of human carcinogens – part D: radiation. Lancet Oncol 10(8):751–752
115. Indoor Tanning Association. Indoor Tanning Association home page. http://www.theita.com/
116. Cokkinides VE, Weinstock MA, O'Connell MC et al (2002) Use of indoor tanning sunlamps by US youth, ages 11–18 years, and by their parent or guardian caregivers: prevalence and correlates. Pediatrics 109(6):1124–1130
117. Hoerster KD, Garrow RL, Mayer JA et al (2009) Density of indoor tanning facilities in 116 large U.S. cities. Am J Prev Med 36(3):243–246
118. Looking Fit (2010) Fact book: the state of the industry report. 2008-2009. http://www.lookingfit.com/articles/fb08a3,p5.html. Accessed 30 July 2010
119. Cho H, Hall JG, Kosmoski C et al (2010) Tanning, skin cancer risk, and prevention: a content analysis of eight popular magazines that target female readers, 1997–2006. Health Commun 25(1):1–10
120. Bellafante G (2001) The way we live now: tan is the new black. The New York Times Magazine 12 Aug 2001
121. Urban Dictionary.com. http://www.urbandictionary.com/define.php?term=Fake%20Bake. Accessed 24 Mar 2010
122. Urban Dictionary.com. http://www.urbandictionary.com/define.php?term=permatan. Accessed 28 Oct 2010
123. Urban Dictionary.com. http://www.urbandictionary.com/define.php?term=tanorexia. Accessed 23 Mar 2010
124. Muther C (2009) You're not bronzed; That tan is fake! The Boston Globe May 21, 2009. http://www.boston.com/ae/celebrity/articles/2009/05/21/youre_not_bronzed_that_tan_is_fake/. Accessed 5 Apr 2010

125. Knight JM, Kirinich AN, Farmer ER et al (2002) Awareness of the risks of tanning lamps does not influence behavior among college students. Arch Dermatol 138(10):1311–1315
126. Hoegh HJ, Davis BD, Manthe AF (1999) Sun avoidance practices among non-Hispanic white Californians. Health Educ Behav 26(3):360–368
127. Mahler HI, Kulik JA, Gerrard M et al (2007) Long-term effects of appearance-based interventions on sun protection behaviors. Health Psychol 26(3):350–360
128. Clarke L, Korotchenko A (2009) Older women and suntanning: the negotiation of health and appearance risks. Sociol Health Ill 31(5):748–761
129. O'Riordan DL, Field AE, Geller AC et al (2006) Frequent tanning bed use, weight concerns, and other health risk behaviors in adolescent females (United States). Cancer Causes Control 17(5):679–686
130. Shoveller JA, Lovato CY, Young RA et al (2003) Exploring the development of sun-tanning behavior: a grounded theory study of adolescents' decision-making experiences with becoming a sun tanner. Int J Behav Med 10(4):299–314
131. Stapleton J, Turrisi R, Todaro A et al (2009) Objectification theory and our understanding of indoor tanning. Arch Dermatol 145(9):1059–1060
132. Greenman J, Jones DA (2010) Comparison of advertising strategies between the indoor tanning and tobacco industries. J Am Acad Dermatol 62(4):685 e681–618
133. Mayer JA, Hoerster KD, Pichon LC et al (2008) Enforcement of state indoor tanning laws in the United States. Prev Chronic Dis 5(4):A125
134. Pichon LC, Mayer JA, Hoerster KD et al (2009) Youth access to artificial UV radiation exposure: practices of 3647 US indoor tanning facilities. Arch Dermatol 145(9):997–1002
135. Demierre MF (2003) Time for the national legislation of indoor tanning to protect minors. Arch Dermatol 139(4):520–524
136. US Food and Drug Administration (2010) Executive summary: the potential risks and benefits of tanning lamps. http://www.fda.gov/downloads/AdvisoryCommittees/CommitteesMeetingMaterials/MedicalDevices/MedicalDevicesAdvisoryCommittee/GeneralandPlasticSurgeryDevicesPanel/UCM205687.pdf. Accessed 30 July 2010
137. Lazovich D, Vogel RI, Berwick M et al (2010) Indoor tanning and risk of melanoma: a case-control study in a highly exposed population. Cancer Epidemiol Biomarkers Prev 19(6):1557–1568

Chapter 3
Prevalence and Correlates of Indoor Tanning

Elliot J. Coups and L. Alison Phillips

Abstract This chapter systematically reviews recent research on the prevalence and correlates of indoor tanning. We review the literature on the extent to which indoor tanning facilities are accessible to individuals in various geographic regions in the United States and internationally. Documenting the prevalence and accessibility of indoor tanning provides an indication of the need for interventions to reduce engagement in this health-damaging behavior, and also facilitates tracking of future trends in indoor tanning. Examination of the correlates of indoor tanning provides insight on populations that might benefit most from relevant policy changes or public health interventions. The chapter also considers several issues regarding the measurement of indoor tanning behaviors. Directions for future research are outlined throughout the chapter.

Keywords Historical context of indoor tanning • Medical effects of indoor tanning • Prevalence of indoor tanning • Correlates of indoor tanning • Availability of indoor tanning • Indoor tanning facilities • Artificial UV exposure • Artificial tanning • Systematic review • Demographic correlates of indoor tanning • Medical history and physical characteristic correlates of indoor tanning • Behavioral correlates of

E.J. Coups (✉)
The Cancer Institute of New Jersey, New Brunswick, NJ, USA

Department of Medicine, UMDNJ-Robert Wood Johnson Medical School, New Brunswick, NJ, USA

Department of Health Education and Behavioral Science, UMDNJ-School of Public Health, Piscataway, NJ, USA
e-mail: coupsej@umdnj.edu

L.A. Phillips
Institute for Health, Health Care Policy, and Aging Research, Rutgers—The State University of New Jersey, New Brunswick, NJ, USA

C.J. Heckman and S.L. Manne (eds.), *Shedding Light on Indoor Tanning*,
DOI 10.1007/978-94-007-2048-0_3, © Springer Science+Business Media B.V. 2012

indoor tanning • Social factors in indoor tanning • Sex differences in indoor tanning • Age differences in indoor tanning • Measurement of indoor tanning

Abbreviations

BMI	Body mass index
IARC	International Agency for Research on Cancer
SAD	Seasonal affective disorder
UV	Ultraviolet
UVB	Ultraviolet B
UVA	Ultraviolet A
US	United States

In this chapter, we present an overview of research on the prevalence and correlates of indoor tanning. We use the term indoor tanning to refer to the use of all types of ultraviolet (UV) light tanning devices, commonly referred to as tanning beds, tanning booths, sunbeds, or solaria. We do not focus on the use of UV-based therapies for medical purposes, such as the treatment of psoriasis or eczema. Also, we do not consider the use of non-UV sunless tanning products and procedures, such as self-applied tanning lotions or creams and professionally-applied spray-on or mist tans, which are addressed by Pagoto in Chap. 11 of this book.

A detailed consideration of the prevalence and correlates of indoor tanning is important for several reasons. Establishing the prevalence of indoor tanning serves to document the extent to which individuals currently engage in this health risk behavior, which indicates the need for interventions to reduce its prevalence. It also provides a benchmark from which to examine potential future changes in the prevalence of indoor tanning, which may be attributable to a variety of factors, including sociocultural trends, enactment or enforcement of regional or national policies, and public health interventions. Identifying correlates of indoor tanning helps to determine groups of individuals who are more likely to engage in indoor tanning and for whom policy changes or public health interventions may be most needed. It also provides insight on individuals' motivations for indoor tanning and thus informs the content of such interventions. In this chapter, we focus on correlates of indoor tanning that are generally not included in common theories of health behavior, which typically emphasize attitudinal, affective and cognitive factors. Thus, the correlates we examine include demographic factors, medical history and physical characteristics, engagement in other health-related behaviors, and social factors such as parental engagement in indoor tanning. Theory-driven motivations for indoor tanning are examined in detail by Hillhouse and Turrisi in Chap. 4 of this book.

We begin the chapter by briefly reviewing the historical and medical contexts of indoor tanning. We then examine the pertinent literature on the availability of indoor tanning facilities. The subsequent two sections focus on the prevalence and correlates of indoor tanning, respectively. Next, we consider measurement issues related to indoor tanning. We end by outlining key points and conclusions.

3.1 Historical and Medical Contexts of Indoor Tanning

An understanding of the history and health effects of indoor tanning provides important contextual information when considering its current prevalence and correlates. We present an overview of relevant issues here; for more information, see several other chapters in this book, including Chap. 2 by Hunt and colleagues and Chap. 6 by Zook and colleagues.

The precursor to the modern-day tanning bed or booth is the sunlamp (or UV lamp), which was in common use in the United States by the late 1920s. At the time, the prevailing belief among many medical professionals and the lay public was that exposure to natural or artificial sources of UV light produced multiple preventive health benefits, including reduced incidence of respiratory infections, improved metabolism and circulation, and increased mental activity [1, 2]. During the 1930s and 1940s, the lack of such salutary effects of sunlamps was increasingly recognized by the medical community. Further, due to clear evidence regarding the adverse health effects of sunlamps, the American Medical Association developed a series of guidelines during this time period for the marketing and manufacturing of sunlamps [2]. Use of sunlamps at this time often resulted in a sunburn rather than a tan, as they utilized a high proportion of ultraviolet B (UVB) radiation. Despite the unequivocal evidence regarding the health-damaging effects of sunlamps by this time, they were still in widespread use through the 1970s. In the late 1970s, the development of indoor tanning beds for home and commercial use by Friedrich Wolff heralded the beginning of the modern-day indoor tanning industry. These tanning beds emitted approximately 95% of their UV light as ultraviolet A (UVA) radiation, thereby reducing the likelihood of incurring a sunburn from indoor tanning.

The medical dangers of indoor tanning are well-documented. In addition to premature aging of the skin, adverse effects of indoor tanning include sunburn, photo-induced medication reactions, polymorphous light eruption (skin rash of red, itchy bumps), sunbed lentigines (flat, brown lesions that typically appear on the arms and legs), atypical melanocytic lesions (abnormal looking moles), ocular disorders (including ocular melanoma), and skin cancer [3, 4]. Indoor tanning increases the risk of both non-melanoma skin cancer (basal and squamous cell carcinoma) and melanoma of the skin, the most deadly form of skin cancer [5, 6]. Use of sunbeds prior to the age of 35 years increases the risk for melanoma by 75% [5]. Indoor tanning is characterized by multiple national and international organizations, including the International Agency for Research on Cancer (IARC), as being carcinogenic to humans [3]. In the United States, approximately 3,000 hospital emergency cases each year are attributable to indoor tanning [7]. Results of a national, population-based survey in the United States in 2004 indicated that 58% of 11–18 year olds who reported indoor tanning in the past year had ever experienced a sunburn due to indoor tanning [8]. The reported sunburn rate was even higher (72%) among individuals reporting six or more indoor tanning sessions in the past year, but did not differ according to age, sex, or sun sensitivity.

In general, public awareness of the health risks of indoor tanning is high, even among indoor tanners [9]. Thus, public health interventions that focus solely on increasing knowledge of the dangers of indoor tanning will likely have little impact on its prevalence.

3.2 Availability of Indoor Tanning Facilities

There is increasing recognition that elements of both the built environment (i.e., buildings and neighborhoods created by people) and the social environment can play an important role in promoting or impeding individuals' engagement in a variety of health-related behaviors. Environmental factors are of particular relevance for a health behavior such as indoor tanning that is performed primarily outside of the home in commercial establishments. While consumer demand for indoor tanning services undoubtedly partly determines the location and density of indoor tanning facilities, it is also likely that the availability of such services also contributes to their utilization. In the United States, indoor tanning is available at more than 40,000 businesses and is estimated to be a $5 billion annual industry [10].

Several research studies have examined the prevalence and density of indoor tanning facilities in a variety of regions and countries. In one of the earliest comprehensive studies, Palmer and colleagues documented the number and density of indoor tanning facilities in 1997/1998 in the 20 most populous cities in each of four regions of the United States (Northeast, Midwest, South, and West) [11]. The prevalence of tanning facilities in each region was determined based on the number of entries in the *Yellow Pages* under the heading of "tanning salons." Each city had an average of 50.3 indoor tanning facilities (standard deviation = 34.5), ranging from 4 in Honolulu (Hawaii) to 134 in Minneapolis (Minnesota). The average density of indoor tanning facilities was 13.89 per 100,000 individuals. Results of multivariate analyses indicated that cities with a higher density of facilities had higher proportions of White individuals, a lower average income, and a lower average daily temperature.

In a more recent analysis, Hoerster and colleagues examined the density of indoor tanning facilities in 116 large US cities in 2006 [12]. On average, there were 41.8 facilities in each city (standard deviation = 30.8), which was higher than the average number of Starbucks coffee shops (mean = 19.0) and McDonald's restaurants (mean = 29.6). This suggests that individuals living or working in large cities commonly have easy access to indoor tanning facilities. The average density was 11.8 per 100,000 individuals, and in multivariate analyses, the factors associated with a higher density of indoor tanning facilities were having a higher proportion of White individuals and a lower average daily UV index (the expected amount of UV radiation that will reach the earth's surface at solar noon each day). Differences in methodology make it hard to draw comparisons between the results of this study and that of Palmer and colleagues [11]. It is also of note that the analyses in these studies focused solely on population density of tanning facilities, which does not necessarily correspond with geographic density. For example, in the study by Hoerster and colleagues, New York City had the greatest number of indoor tanning facilities ($N=183$) of all 116 cities [12]. However, given its population of more than eight million individuals, it had the fourth lowest indoor tanning facility density (2.3/100,000 individuals). Consideration of the geographic size of New York City would likely reveal a different picture compared to other cities in terms of the density of facilities per square mile, which provides an indication of the distance a person would need to travel in order to visit an indoor tanning facility. Future research is needed to examine the issue of population density versus geographic density of indoor tanning facilities.

There are also data available regarding the availability of indoor tanning facilities in Australia and New Zealand. A review of *Yellow Pages* telephone directory listings for all 18 regions in New Zealand revealed a 425% increase in the number of indoor tanning wholesale businesses from 1992 to 2006 and a 141% increase in the number of facilities offering indoor tanning [13]. During the same time period, there was a 383% increase in the number of indoor tanning facilities in the eight capital cities in Australia [14], followed by a 32% decrease from 2006 to 2008/2009 [15]. This decrease was attributed largely to negative publicity about indoor tanning and skin cancer and subsequent changes in state legislation after the widely reported death of a 26-year-old woman from melanoma in August 2007. Some indoor tanning facilities reported up to a 75% decrease in revenue during the months after the negative media attention [16]. Among adults in Western Australia, awareness of the elevated cancer risk associated with indoor tanning increased from 71% in September 2006 to 92% in September 2007 [17]. Overall, the findings of these studies suggest that the availability and utilization of indoor tanning facilities are responsive to changes in legislation and public awareness of the adverse effects of indoor tanning. Future research is needed to further examine the direction(s) of causality of consumer demand for, and industry supply of, indoor tanning facilities [12].

3.3 Prevalence of Indoor Tanning

We sought to identify up to date information on the prevalence of indoor tanning. Search strategies included the following: (1) a search of PubMed, PsycInfo, Google Scholar, and Web of Science databases using relevant key terms (e.g., indoor tanning, sunbed*, tanning bed*, tanning booth*, tanning salon*, solarium*, solaria, sunlamp*, artificial tanning, UV tanning) and subject headings; (2) review of reference lists of articles identified for this review; (3) cited reference searches of articles identified for this review; and (4) examination of 16 articles (published from January 2000 to August 2008) included in a recent review of indoor tanning correlates [18].

We included research articles in the current systematic review that met the following criteria: (1) provided data on the prevalence of indoor tanning outside of the context of an intervention; (2) published in a peer-reviewed journal in the English language; (3) published prior to May 2010; and (4) data on indoor tanning prevalence were collected from January 2000 onwards (in order to focus on more recent estimates of indoor tanning prevalence). We did not include studies that focused solely on indoor tanning attitudes, beliefs, intentions, or policy. For several articles, we queried the authors in order to establish the time frame of data collection. For studies that reported data on indoor tanning prevalence at specific time points prior to and after January 2000, we only included the data from 2000 onwards [8, 19, 20]. A study by Bagdasorov and colleagues reported indoor tanning prevalence only in terms of the mean number of occasions per person [21]. The authors provided us with information on the past year and lifetime prevalence of indoor tanning among the study participants.

We located a total of 48 research articles (published from 43 separate data sets) that met our inclusion criteria. Table 3.1 provides information regarding the characteristics of the 48 articles, which are organized by time frame of data collection within geographical regions (the Americas, Europe, and Australia). Articles that were published on a single data set are included together in the table. Most of the articles focused on data from the United States, with the remainder being conducted primarily in one or more European countries or regions. The sample sizes ranged from 99 beachgoers in Texas [27] to a national probability sample of 29,394 US adults [37]. Around half of the studies focused exclusively on younger-aged populations including adolescents and high school students, university students, or adults under the age of 30 years. This attention is warranted, given the higher prevalence of indoor tanning among these groups compared to adults over the age of 30 years [37]. Convenience sampling was used in many studies of younger-aged populations (e.g., university students taking a specific course), although several did use population-based sampling methodologies [24, 29, 40, 62]. In contrast, most of the studies that did not focus exclusively on younger individuals utilized population-based sampling, which provides more generalizable estimates of indoor tanning prevalence than convenience sampling.

A majority of the studies shown in Table 3.1 reported either past-year or lifetime prevalence of indoor tanning. There are several notable findings evident from the table. First, the prevalence of indoor tanning varied considerably according to the age group of focus. Among population-based samples that focused primarily on adults, the prevalence of indoor tanning in the past year (from lowest to highest) was 1% in Queensland, Australia [65], from 5% to 14% in the United States [37, 38, 41], 21% in Mannheim, Germany [60, 61], and 30% in Denmark [58]. Among university student samples in the United States, the past year prevalence of indoor tanning varied from 33% to 60% [21, 45]. Four studies in the United States examined rates of indoor tanning during the past year among adolescents (aged 11–18 years), with a rate of 11% reported in two national studies and a study conducted in Miami (Florida) [8, 24, 40, 44] and a rate of 26% in a study conducted in Minneapolis-St. Paul (Minnesota) and Boston (Massachusetts) [24]. Lifetime rates of indoor tanning among adolescents were reported in studies from multiple countries (listed from lowest to highest): 2–4% among high school and university students in Brazil [29, 49];

Table 3.1 Prevalence of indoor tanning

References	Study region	Data collection Time frame	*N*	Age (years) (*M* = mean)	Sample	Prevalence of indoor tanning		
						Past year	Lifetime	Other
The Americas								
Clough-Gorr et al. [22]	US, New Hampshire	1996–2001	423 Melanoma patients; 678 controls	20–69	Melanoma cases from the New Hampshire cancer registry and a randomly selected control sample of age- and sex-matched licensed drivers	–	Sunlamp or tanning bed: 37.1% of patients, 32.0% of controls; Sunlamp: 17.1% overall (20.3% of patients, 15.0% of controls); Tanning bed: 21.9% overall (22.9% of patients, 20.9% of controls)	–
Lucci et al. [23]	US, Dallas and Houston (Texas)	November 1999–February 2000	210	14–19	All students taking a mandatory health education class	–	–	18% in the past 6 months
Lazovich et al. [24]; Stryker et al. [25]; Zeller et al. [26]	US, Minneapolis-St. Paul (Minnesota) and Boston (Massachusetts)	Fall 2000	1,273	14–17	List of individuals obtained from a commercial list targeted by age	25.8% overall; 36.7% of girls; 7.7% of boys	30.3% overall; 41.6% of girls; 12.0% of boys	30% of those who tanned more than once in the past year had difficulty quitting
Ramirez et al. [27]	US, Galveston Island (Texas)	Summer 2001	99	18–61 (*M* = 29)	Beachgoers	–	43.4%	19.2% ≥4 times/month

(continued)

Table 3.1 (continued)

References	Study region	Data collection Time frame	N	Age (years) (M = mean)	Sample	Prevalence of indoor tanning		
						Past year	Lifetime	Other
Lazovich et al. [28]	US, Minnesota	2002	802	≥18	Randomly selected households in Minnesota	–	38.0% overall; 44.9% of women; 30.2% of men	–
Benvenuto-Andrade et al. [29]	Brazil, Porto Alegre	September–November 2002	724	12–19 (M = 16.5)	Students from 25 randomly selected schools (19 public, 6 private)	–	3.5% overall; 5.0% of girls; 1.6% of boys	–
Buchanan et al. [30]	US and Canada	2003	9,232 cancer survivors; 2,931 siblings	17–54 (M = 31 for survivors)	Retrospective cohort study of ≥5-year survivors of childhood or adolescent cancer and sibling controls	15.0% of cancer survivors; 19.7% of siblings	33.3% of cancer survivors; 44.5% of siblings	–
Danoff-Burg and Mosher [31]	US, Albany (New York)	November–December 2003	164	18–23	Undergraduate students taking introductory psychology	35% overall; 56.8% of women; 15.9% of men	–	–
Hillhouse et al. [32]	US, Johnson City (Tennessee)	November 2003–March 2004	126	College-aged	Female undergraduate students	–	–	34.1% regularly tanned; 21.4% frequently tanned
Sheehan and Lesher [33]	US, Augusta (Georgia)	February–March 2004	121	14–58 (M = 31.3)	Individuals receiving spray-on sunless tanning at an indoor tanning salon	–	86.0%	–

Cokkinides et al. [8]; Bandi et al. [19]	US	2004	1,589 adolescents; 1,931 parents/ guardians	Adolescents: 11–18; parents: 27 to >46	National, population-based sample	Adolescents: 11.1% overall; Parents: 12.8% overall; 16.0% of 27–45 years; 7.2% of ≥46 years; 14.8% of women; 5.3% of men	–	–
Ting et al. [34]	US, Iowa City (Iowa)	2004	1,518	<18 to >60	Patients attending a dermatology clinic	–	32.1%	–
Johnson et al. [35]	US	September–November 2004	364	<30 to >70	Random selection of physicians from four specialties from American Medical Association records	2.8% overall; 4.5% of women; 1.8% of men	18.6% overall (6% reported medical use); 26.1% of women; 14.4% of men; 16% of dermatologists; 21% of non-dermatologists	–
Brooks et al. [36]	US, Boston (Massachusetts)	July 2004	448	18–30	Sample of individuals in universities, shopping venues, and parks	33%	–	–

(continued)

Table 3.1 (continued)

References	Study region	Data collection Time frame	*N*	Age (years) (*M* = mean)	Sample	Prevalence of indoor tanning: Past year	Lifetime	Other
Heckman et al. [37]; Coups et al. [38]	US	2005	29,394 (Heckman); 28,235 (Coups)	≥18	National, population-based sample	13.8% overall; 20.4% of 18–29 years (27.3% of women, 13.2% of men); 16.7% of 30–39 years (21.9% of women, 11.6% of men); 13.6% of 40–49 years (15.7% of women, 11.3% of men); 9.8% of 50–64 years (10.3% of women, 9.3% of men); 7.8% of ≥65 years (7.3% of women, 8.4% of men)	–	–
Stapleton et al. [39]	US, State College (Pennsylvania)	2005 (fall and winter months)	174	*M* = 19.83	Undergraduate students taking an introductory health course, excluding skin types IV and V	42.9%	–	–

Hoerster et al. [40]	US (in the 100 largest cities)	January–December 2005	5274	14–17	Households targeted based on age and drawn from phone directories cross-referenced with secondary sources	11.2% overall; 6.2% of 14 years; 8.8% of 15 years; 13.0% of 16 years; 17.5% of 17 years; 18.1% of girls; 3.5% of boys	–	–
Stryker et al. [41]	US	February–August 2005	5,491	≥18	National, population-based sample	11.3% of women 5.1% of men	–	–
Cafri et al. [42]	US, Tampa (Florida)	October–November 2005	311	18–26	Female undergraduate students	–	–	39.3% in the past 6 months
Bagdasarov et al. [21]	US, New Brunswick (New Jersey)	December 2005	745	19–25	Undergraduate students (with data excluded from the 130 students reporting the darkest skin color)	33.0%	46.0%	–
Poorsattar and Hornung [43]	US, Seattle (Washington)	December 2005–January 2006	355	17–30	Undergraduate students	21.9%	33.1% overall; 42% of women; 17% of men;	28% of indoor tanners met criteria for UV light substance-related disorder
Ma et al. [44]	US, Miami (Florida)	2006	369	$M = 16.9$	High school students	11.5% overall; 15.1% of White Hispanics; 6.1% of White Non-Hispanics	–	–

(continued)

Table 3.1 (continued)

References	Study region	Data collection Time frame	N	Age (years) (M = mean)	Sample	Prevalence of indoor tanning		
						Past year	Lifetime	Other
Hillhouse et al. [45]	US, Johnson City (Tennessee)	January–April 2006	853	18–54 (M = 21.8)	Undergraduate students	60%	–	–
Heckman et al. [46]	US, Richmond (Virginia)	Spring 2006	400	17–30	College-aged students and members of the community	–	38.3%	26.5% of the total sample and 44.4% of tanners had tanning dependence
Mosher and Danoff-Burg [47]	US, New York	September–December 2006	421	college-aged (74% 18–19 years)	Undergraduate students from a psychology research participant pool	–	56.3%	30.6–39.3% of indoor tanners met criteria for tanning addiction
Gordon and Guenther [48]	Canada, London (Ontario)	November–December 2006	1,202	Tenth grade	Tenth grade students from eight schools chosen to be representative of London (Ontario)	–	14%	3.7% ≥ once/week
Castilho et al. [49]	Brazil, Taguatinga (Brasilia)	February–April 2007	368	M = 22.1	University students taking health sciences or human sciences courses	–	1.9% overall; 3.5% of women; 0.0% of men	–
Robinson et al. [20]	US, Chicago (Illinois)	2007	100	18–30 (estimated)	Beachgoers who were age- and sex-matched to random-digit-dialed participants from 1988 to 1994 surveys	27%	–	–

Europe								
Bataille et al. [50]	Belgium, France, the Netherlands, Sweden, United Kingdom	1999–2001	597 melanoma patients; 622 controls	18–49	Melanoma cases and age- and sex-matched controls	–	Melanoma cases: 52.8% overall; 61% of women; 34% of men; Controls: 56.9% overall; 61% of women; 49% of men	–
Ezzedine et al. [51]	France	2001	7,303	Women: 35–60; men: 45–60	Participants in a longitudinal cohort study, recruited using a national multimedia campaign with selection according to sex, age group, smoking habits, and geographical location	–	14.9% overall; 20.9% of women; 6.4% of men	–
Bränström et al. [52]	Sweden, Stockholm	May 2001	1,752	18–37	Random, population-based sample stratified by age and sex	–	–	35.4% current use
Devos et al. [53]	Belgium, Ostend	May–August 2001	360	0 to >60	Beachgoers	–	–	21.9% current use ("from time to time"); 30.2% of females; 9.7% of males

(continued)

Table 3.1 (continued)

References	Study region	Data collection Time frame	*N*	Age (years) (*M* = mean)	Sample	Prevalence of indoor tanning		
						Past year	Lifetime	Other
Dissel et al. [54]	Germany, North-Rhine Westphalia	2002–2006	1,242	<29 to >64	Individuals undergoing skin cancer screening	–	–	15.4% regular use (≥11 times/ year); 29.5% current use (≥1 time/ year); Current use: 50.0% of <29 years; 27.9% of 26–64 years; 11.5% of ≥65 years; 37.8% of females; 23.6% of males
Hamlet and Kennedy [55]	United Kingdom, Lanarkshire (Scotland)	June 2003	1,405	8–11	Students at 23 schools; school nurses counted a show of hands in response to each question asked	–	–	6.8% in the past 5½ months; 9.9% of 8–9 years; 5.5% of 10–11 years
Mackay et al. [56]	United Kingdom, Merseyside (England)	Autumn 2004	496	14–16	Children in two schools representing urban, relatively deprived regions	–	42.9% overall; 60% of girls; 25% of boys;	–

Börner et al. [57]	Germany	2007	1,501	≥14	National, population-based survey	–	28.4% overall; 33.6% of women; 22.8% of men; 18.5% of 14–17 years; 39.1% of 18–29 years; 43.2% of 30–44 years; 22.0% of 45–59 years; 14.8% of≥60 years	39.7% of tanners were "frequent tanners" (≥10 times/year)
Køster et al. [58]	Denmark	March 2007	3,437	15–59	National, population-based sample	28.2% overall; 36% of females; 18% of males	61.5%	15.8% of total sample ≥ once/month
Roberts and Foley [59]	United Kingdom, South Wales	2008	203	14–15	Children in two schools selected to represent two distinct geographical areas	–	21.7% overall; 32.1% of girls; 9.6% of boys	5.5% of girls weekly use
Diehl et al. [60]; Schneider et al. [61]	Germany, Mannheim	2008	500	18–45	Population-based sample	21.0% overall; 26.2% of 18–25 years; 15.1% of 26–35 years; 22.9% of 36–45 years; 26.6% of women; 16.0% of men	46.7% overall; 41.5% of 18–25 years; 48.6% of 26–35 years; 48.4% of 36–45 years; 59.0% of women; 34.8% of men	–

(continued)

Table 3.1 (continued)

References	Study region	Data collection Time frame	N	Age (years) (M=mean)	Sample	Prevalence of indoor tanning		
						Past year	Lifetime	Other
Thomson et al. [62]	United Kingdom, England	2008–2009	6,209 (Six Cities Study); 3,101 (National Prevalence Study)	11–17	Population-based sample in six cities in England (Six Cities Study) and nationally in England (National Prevalence Study)	–	Six Cities Study: 10.8% overall; 4.3% of 11–14 years; 19.5% of 15–17 years; 14.4% of girls; 7.3% of boys; National Prevalence Study: 6.0% overall; 1.8% of 11–14 years; 11.2% of 15–17 years; 8.6% of girls; 3.5% of boys	–
Australia								
Paul et al. [63]	Australia, New South Wales	2000	1,509	≥15	Random sample of residents	–	10% overall; 13.7% of women; 4.8% of men; 3.9% of 15–17 years; 11.1% of 18–29 years; 18.5% of 30–39 years;	–

Dobbinson et al. [64]	Australia	November 2003–January 2004 (summer)	699 adolescents; 5,073 adults	12–17; 18–69	National, population-based sample, stratified by age, sex, and area of residence	–	3% of adolescents (7% of girls, 15–17 years); 11% of adults (14% of women, 25–44 years)	–
Lawler et al. [65]	Australia, Queensland	February–November 2004	9,298	20–75	Random sample of households stratified by age, sex, and geographic region	1.3%	10.7%	–

3% among 12–17 year-olds in Australia [64]; 6–11% among 11–17 year-olds in the United Kingdom [62]; 14% among tenth grade students in London, Ontario (Canada) [48]; 22% among 14–15 year-olds in South Wales [59]; 30% among 14–17 year-olds in Minneapolis-St. Paul (Minnesota) and Boston (Massachusetts) [26]; and 43% among 14–16 year-olds in Merseyside, England [56].

A second notable finding from Table 3.1 is that indoor tanning prevalence varied according to the population group studied, including among beachgoers [20, 27, 53], individuals undergoing spray-on sunless tanning at an indoor tanning salon [33], individuals diagnosed with melanoma [22, 50], survivors of childhood or adolescent cancer and their siblings [30], and physicians (dermatologists and non-dermatologists) [35]. Among individuals receiving a spray-on sunless tan at a tanning salon, 86% reported ever having tanned indoors, suggesting that these two types of tanning often co-occur [36, 41, 46, 48]. Survivors of childhood or adolescent cancer reported lower rates of past year and lifetime indoor tanning than their siblings (past year: 15% vs. 20%; lifetime: 33% vs. 45%) [30]. In a national survey of dermatologists, internists, family medicine physicians, and pediatricians in the United States, fewer than 3% reported indoor tanning in the past year, and 13% reported ever having done so (an additional 6% reported ever having tanned indoors for medical reasons) [35].

A third noteworthy finding from Table 3.1 is that there is evidence from several studies in the United States that some adolescents and young adults may be addicted to, or dependent on, indoor tanning. Zeller and colleagues noted that 30% of indoor tanners in their study reported that they would have difficulty quitting [26]. Estimated rates of tanning dependence among college student indoor tanners have varied from 28% to 44% [43, 46, 47]. The criteria used to determine tanning dependence have varied across research studies, and further research is needed with regard to measurement of dependence, as well as to identify potential underlying biological mechanisms of tanning dependence. For more information on the topic of indoor tanning dependence, see Chap. 7 by Shah and Feldman in this book.

As outlined in the preceding discussion, prevalence rates of indoor tanning vary widely across research studies. This variation is partly due to differences in study region, sampling approaches, and measurement of indoor tanning (see Sect. 3.5), but also largely reflects variation in the demographic characteristics—most notably, age and sex—of study participants. This raises the important issue of correlates of indoor tanning, which is addressed in detail in the next section.

3.4 Correlates of Indoor Tanning

We examined the correlates of indoor tanning reported in each of the studies shown in Table 3.1. As noted earlier, for the purposes of this chapter, we did not consider correlates of indoor tanning that are typically included in health behavior theories (see Chap. 4 in this book). Table 3.2 shows the nature of the association (i.e., statistically significant positive, statistically significant negative, or no statistically significant association) between each correlate and indoor tanning for each of the

Table 3.2 Correlates of indoor tanning

Correlates	Significant positive association	Significant negative association	Association not significant
Demographic factors			
Age	All studies shown in Table 3.1 that examined the association between age and indoor tanning (except Bagdasarov et al. [21], Bataille et al. [50], Ezzedine et al. [51] among men, and Schneider et al. [61]) found it to be statistically significant. Overall, there is evidence for a curvilinear association between age and indoor tanning, such that the prevalence of indoor tanning is low in early adolescence and increases with age up until the late teens or early-to-mid twenties, after which it begins to decline.		
Female sex	Bagdasarov et al. [21] Bataille et al. [50] Benvenuto-Andrade et al. [29] Börner et al. [57] Bränström et al. [52] Castilho et al. [49] Cokkinides et al. [8] and Bandi et al. [19] Devos et al. [53][a] Diehl et al. [60] and Schneider et al. [61] Dissel et al. [54] Ezzedine et al. [51] Gordon and Guenther [48] Heckman et al. [37] (18–49 years) Hoerster et al. [40] and Johnson et al. [35] (ever use) Lawler et al. [65] Lazovich et al. [24] and Stryker et al. [25] Lazovich et al. [28] Mackay et al. [56] Poorsattar and Hornung [43] Thomson et al. [62]	–	Dissel et al. [54] (frequency of use) Ezzedine et al. [51] (frequency of use) Heckman et al. [37] (≥50 years) Heckman et al. [46] (tanning dependency vs. no dependency) Johnson et al. [35] (past year use) Mosher and Danoff-Burg [47] (tanning dependency vs. no dependency)

(continued)

Table 3.2 (continued)

Correlates	Significant positive association	Significant negative association	Association not significant
Caucasian race	Bandi et al. [19] Gordon and Guenther [48][b] Heckman et al. [37] (18–49 years) Heckman et al. [46] (tanning dependency vs. no dependency) Hoerster et al. [40] Lazovich et al. [28] (women) Stryker et al. [41]	–	Heckman et al. [37] (≥50 years) Lazovich et al. [28] (men) Ma et al. [44]
Education level	Börner et al. [57] Diehl et al. [60] and Schneider et al. [61] (moderate amount of education vs. low or high education) Dissel et al. [54] Heckman et al. [37] (18–39 years) Lazovich et al. [28] (men) Stryker et al. [41]	Bränström et al. [52] Køster et al. [58]	Bandi et al. [19] Heckman et al. [37] (≥40 years) Lazovich et al. [28] (women)
Single marital status	Køster et al. [58] (men)	Börner et al. [57] (indoor tanners less likely to be widowed)	Køster et al. [58] (women) Schneider et al. [61]
Current employment	Diehl et al. [60] and Schneider et al. [61] (current use) Dissel et al. [54] Ezzedine et al. [51] (women) Lawler et al. [65]	–	Ezzedine et al. [51] (men) Schneider et al. [61] (ever use)
Outdoor occupation	–	Ezzedine et al. [51] (women)	Ezzedine et al. [51] (men)

Region of residence (see Sect. 3.4.1 in the text for details regarding the association with indoor tanning)	Bataille et al. [50] Ezzedine et al. [51] (women) Heckman et al. [37] (18–29 years[b]) Lawler et al. [65] Lazovich et al. [24] and Stryker et al. [25] Stryker et al. [41] Thomson et al. [62]	–	Cokkinides et al. [8] Ezzedine et al. [51] (men) Heckman et al. [37] (≥30 years)
Household income (or social grade)	Lawler et al. [65] Lazovich et al. [28] (men) Stryker et al. [41]	Thomson et al. [62]	Lazovich et al. [28] (women)
Medical history and physical characteristics			
Personal cancer history	–	Buchanan et al. [30]	Bataille et al. [50] (melanoma cases vs. controls)
Family cancer history	–	–	Heckman et al. [37] Køster et al. [58] Schneider et al. [61]
Seasonal affective disorder	Hillhouse et al. [32] (women)	–	–
Ultraviolet (UV) light disorder (addiction)	Poorsattar and Hornung [43]	–	–
Skin sensitivity (including skin color, suntan intensity, and sensitivity/reactivity to UV exposure)	Ezzedine et al. [51] (fair natural skin color, among women) Ezzedine et al. [51] (skin sensitivity) Heckman et al. [37] (18–49 years for skin reaction after 1 h in the sun[b]) Heckman et al. [46] (tanning dependency vs. no dependency)	Bandi et al. [19] Bataille et al. [50] Bränström et al. [52] Diehl et al. [60] and Schneider et al. [61] (current use) Dissel et al. [54] Heckman et al. [37] (18–29 years for skin reaction after 2 weeks in the sun) Hoerster et al. [40] Køster et al. [58]	Ezzedine et al. [51] (fair natural skin color, among men) Heckman et al. [37] (≥30 years for skin reaction after 2 weeks in the sun) Heckman et al. [37] (≥50 years for skin reaction after 1 h in the sun) Lawler et al. [65] Mosher and Danoff-Burg [47] (tanning dependency vs. no dependency) Schneider et al. [61] (ever use) Stryker et al. [25]

(continued)

Table 3.2 (continued)

Correlates	Significant positive association	Significant negative association	Association not significant
Eye color	–	–	Dissel et al. [54]
Presence of freckles	Ezzedine et al. [51]	–	–
Number of moles	–	–	Dissel et al. [54] Schneider et al. [61]
Dark hair color	–	Ezzedine et al. [51] (women)	Dissel et al. [54] Ezzedine et al. [51] (men)
Behavioral factors			
Stay in shade when outdoors	–	Heckman et al. [37] (18–64 years; 18–29 years[b] and 50–64 years[b])	Heckman et al. [37] (≥65 years)
Sun exposure and sunbathing	Ezzedine et al. [51] (hobbies, among women) Ezzedine et al. [51] (lifetime exposure, among women) Ezzedine et al. [51] (mountain sports, women[b]) Ezzedine et al. [51] (nautical sports) Heckman et al. [46] (hours of sunbathing; tanning dependency vs. no dependency) Køster et al. [58] Ramirez et al. [27]	Bagdasarov et al. [21][b]	Ezzedine et al. [51] (hobbies, among men) Ezzedine et al. [51] (lifetime exposure, among men) Heckman et al. [46] (hours of incidental sun exposure; tanning dependency vs. no dependency)
Frequency of sunscreen use	Ezzedine et al. [51] Heckman et al. [37] (18–49 years[b])	Lazovich et al. [24] Ramirez et al. [27]	Heckman et al. [37] (≥50 years) Stryker el al. [41]
Frequency of wearing protective clothing in the sun	–	Ezzedine et al. [51] (women) Heckman et al. [37] (18–64 years, varied according to the type of clothing[b]) Stryker et al. [41]	Ezzedine et al. [51] (men) Heckman et al. [37] (≥65 years; ≥30 years, varied according to the type of clothing)

Use of sunglasses	Ezzedine et al. [51] (women)	–	Ezzedine et al. [51] (men)
Frequency of sunburn occurrence	Ezzedine et al. [51] (during adulthood, among women) Heckman et al. [37] Heckman et al. [46] (past year sunburns; tanning dependency vs. no dependency) Køster et al. [58]	–	Ezzedine et al. [51] (during adulthood, among men) Ezzedine et al. [51] (during childhood) Heckman et al. [46] (lifetime sunburns; tanning dependency vs. no dependency) Schneider et al. [61] (during childhood)
Use of sunless tanning products	Brooks et al. [36] Gordon and Guenther [48] Heckman et al. [46] (tanning dependency vs. no dependency)	–	–
Ever having a total skin examination	–	–	Heckman et al. [37]
Physical activity	Diehl et al. [60] (current use) and Schneider et al. [61] (ever use) (individual sports vs. team or no sports) Heckman et al. [37] (18–29 years[b] and 40–49 years[b]) Heckman et al. [46] (days of anaerobic activity; tanning dependency vs. no dependency)	–	Heckman et al. [37] (30–39 years and ≥50 years) Heckman et al. [46] (days of aerobic activity; tanning dependency vs. no dependency) Schneider et al. [61] (current use)
Fruit and vegetable intake	–	Heckman et al. [37] (18–49 years[b])	Heckman et al. [37] (≥50 years)
Body mass index	–	Heckman et al. [37] (30–39 years[b]) Heckman et al. [46] (tanning dependency vs. no dependency)	Heckman et al. [37] (18–29 years, ≥40 years)
Alcohol intake	Bagdasarov et al. [21] Heckman et al. [37] (18–29 years[b]) Mosher and Danoff-Burg [47] (non-addicted vs. addictive tendencies and addicted tanners)	–	Heckman et al. [37] (≥30 years)

(continued)

Table 3.2 (continued)

Correlates	Significant positive association	Significant negative association	Association not significant
Smoking	Diehl et al. [60] and Schneider et al. [61] (current use) Ezzedine et al. [51] (women) Heckman et al. [37] (18–29 years[b]) Heckman et al. [46] (days smoked in last 30; tanning dependency vs. no dependency) Lazovich et al. [24] Mackay et al. [56]	–	Bagdasarov et al. [21] Ezzedine et al. [51] (men) Heckman et al. [37] (≥30 years) Heckman et al. [46] (number of cigarettes/day; tanning dependency vs. no dependency) Mosher and Danoff-Burg [47] (non-addicted vs. addictive tendencies and addicted tanners) Schneider et al. [61] (ever use)
Substance use (other than alcohol and tobacco)	Mosher and Danoff-Burg [47] (marijuana use and total number of non-alcohol substances used; non-addicted vs. addictive tendencies and addicted tanners)	–	Mosher and Danoff-Burg [47] (use of stimulants only; non-addicted vs. addictive tendencies and addicted tanners)
Visited physician in past year	Heckman et al. [37] (18–39 years[b])	–	Heckman et al. [37] (≥40 years)
Social factors			
Parents' age	Hoerster et al. [40]	Stryker et al. [25][a]	–
Parents' education	–	Hoerster et al. [40][b] Stryker et al. [25][a]	–
Parental permission for indoor tanning	Cokkinides et al. [8] Hoerster et al. [40] Lazovich et al. [24] and Stryker et al. [25]	–	–
Parents' concern about child tanning indoors	–	Stryker et al. [25]	–

Parents' indoor tanning	Cokkinides et al. [8]	–	–
	Gordon and Guenther [48]		
	Hoester et al. [40]		
	Stryker et al. [25]		
Family members' indoor tanning	Gordon and Guenther [48]	–	–
	Mackay et al. [56]		
Friends' and acquaintances' indoor tanning	Bagdasarov et al. [21] (acquaintances)	–	–
	Bagdasarov et al. [21] (friends[b])		
	Bränström et al. [52]		
	Gordon and Guenther [48]		
	Hoerster et al. [40]		
	Lazovich et al. [24]		
Restrictive state legislation	–	–	Cokkinides et al. [8]

[a]An analysis to determine the statistical significance of this association was conducted by the current authors based on the results presented in the original article

[b]Not statistically significant in a multivariate analysis

research studies. When a study reported both the bivariate and multivariate association between a correlate and indoor tanning, we report in the table all instances when a statistically significant bivariate associate was subsequently found not to be statistically significant in the multivariate analysis. Below we highlight key findings shown in Table 3.2.

3.4.1 Demographic Factors

3.4.1.1 Sex and Age

By far the most frequently assessed correlates of indoor tanning behavior are sex and age. Almost all studies that examined potential sex differences found a higher prevalence of indoor tanning among girls and women compared to boys and men. In a large, population-based study, Heckman and colleagues did not find sex differences in the prevalence of indoor tanning among individuals aged 50 years and older, which may in part be due to the relatively low overall indoor tanning rates among these individuals [37]. Two studies of French and German adults did not find any difference in the frequency of indoor tanning between female and male indoor tanners [51, 54]. A study of US physicians found that more women than men had ever tanned indoors, but there was no difference with regard to the prevalence of indoor tanning in the past year [35]. Two additional studies of US college students reported no difference in the rate with which female and male indoor tanners were categorized as having tanning dependence [46, 47]. Overall, these results indicate that females are more likely to engage in indoor tanning than males, but among indoor tanners, the groups may not differ with regard to the number of tanning occasions or their experience of tanning dependence.

Age does not have a simple, linear relationship with indoor tanning behaviors. The overall picture that emerges across research studies indicates a curvilinear relationship between age and indoor tanning. The prevalence of indoor tanning is typically low among young adolescents and increases with age up until the late teens or early-to-mid 20s, after which it begins to decline. However, it is not possible to rule out the possibility of a cohort effect, as there is a lack of large-scale longitudinal studies tracking engagement in indoor tanning across the lifespan.

3.4.1.2 Race, Education, and Income

There is consistent evidence that the prevalence of indoor tanning is higher among Caucasian individuals compared to other racial groups. In line with this finding, several studies have limited their focus to Caucasian samples or have excluded individuals reporting a darker skin type [21, 39]. Findings regarding the association between indoor tanning and both education and income have been equivocal. Indoor tanning has been found to be inversely associated with education and income in some studies but positively associated in others [28, 37, 41, 52, 54, 57, 58, 60, 62, 65]. In a study of German adults, the prevalence of indoor tanning was highest among those with a moderate amount of education compared to those with less or more education [60, 61] The potentially complex association between indoor tanning and both education and income level should be examined further in future research. Less-educated individuals may tan indoors partly because they are less aware of the risks of indoor tanning. More-educated individuals may tan indoors in part because they have more money to spend on tanning. Further, group norms of beauty and attitudes on indoor tanning may differ between socioeconomic classes, which are defined largely by education and income.

3.4.1.3 Geographic Region and State Legislation

Regional differences in the prevalence of indoor tanning have been found within and across multiple countries. Results from several studies in the Northern hemisphere indicate that the prevalence of indoor tanning is higher among individuals living at more northerly latitudes [50, 51, 62]. In a sample of US adults, Stryker and colleagues reported the highest rates of indoor tanning among those residing in the Midwest, with rates in the Northeast similar to those in the South and West [41]. Heckman and colleagues reported higher rates of indoor tanning among younger adults (aged 18–29 years) in the Northeast and Midwest regions of the United States compared to the South and West [37]. Of note, physicians in the Northeast and Midwest have been found to be more supportive of indoor tanning as a means of improving mood, treating depression, and preventing vitamin D deficiency than those in the South or West [35]. Physicians who hold such beliefs may be less likely to discuss the risks of indoor tanning with their patients and may even promote indoor tanning as a way to alleviate mood disorders or vitamin D deficiency. Given that physicians are trusted sources of information about indoor tanning, it is important that they provide patients with evidence-based information about its health-damaging effects [20].

Several studies have examined whether the prevalence of indoor tanning is different for individuals residing in urban versus rural areas. In a study conducted in Queensland (Australia), adults living in more remote areas were less likely to engage in indoor tanning than those living in more urban areas [65]; however, Thomson and colleagues found use of sunbeds was lowest in London compared to less urban areas of England, and Zahnd and colleagues reported no differences in indoor tanning prevalence between rural and urban US adults [62, 65, 66]. Identifying differences in the prevalence of indoor tanning across regions is important, because it can guide the need for targeted interventions to reduce indoor tanning rates.

The presence of local legislation regarding minors' use of indoor tanning may affect regional tanning prevalence rates. However, Cokkinides and colleagues found no difference in the prevalence of indoor tanning among US adolescents according to whether the state in which the individual resided had legislation restricting minors' access to indoor tanning [8]. When this study was conducted in 2004, 19 states had such legislation, compared to at least 32 as of April 2010 [67]. The lack of association between indoor tanning prevalence and the presence of relevant state legislation may be due to differential enforcement of such legislation across states. Further research is needed to determine the extent to which the enactment and enforcement of legislation influences the prevalence of indoor tanning among minors.

3.4.2 Medical History and Physical Characteristics

Individuals' personal or family history of cancer or other medical conditions may influence their decision about whether to engage in indoor tanning. In a recent study conducted in the United States, survivors of childhood or adolescent cancer were found to have a lower prevalence of indoor tanning than their siblings [30]. A study conducted in Europe found no difference in the percentage of melanoma patients and matched controls who had ever engaged in indoor tanning [50]. The results of this latter study stand in contrast to the evidence from a recent systematic review that linked indoor tanning with an increased risk for melanoma [5]. A family history of skin cancer or any cancer was not associated with the prevalence of indoor tanning in three studies [37, 58, 61]. With regard to other medical conditions, Hillhouse and colleagues found a positive association between symptoms of seasonal affective disorder (SAD) and frequent indoor tanning (defined as 40 or more times in a year) among female undergraduates [32]. This suggests that some high-frequency indoor tanners may tan in an effort to improve their mood and ameliorate symptoms of SAD. Future research is warranted on this topic, for which it will be important to examine potential psychosocial and biological factors that may underpin frequent indoor tanning or tanning dependency. Of note, Poorsattar and Hornung found that students characterized as having a UV light substance-related disorder were more likely to be indoor tanners than other students [43]. For a detailed discussion of issues related to tanning dependency, see Chap. 7 in this book by Shah and colleagues.

Individuals' physical characteristics, such as the sensitivity of their skin to tanning, may also affect their engagement in indoor tanning. Although the results of relevant studies have been somewhat equivocal, the overall evidence suggests a curvilinear relationship such that indoor tanning may be more prevalent among individuals with low to moderate skin sensitivity to the sun (e.g., Fitzpatrick skin type III or IV) compared to individuals with highly sensitive skin that burns very easily or those with dark or sun insensitive skin. Motivations for indoor tanning may differ according to individuals' skin sensitivity. For example, Ezzedine and colleagues found that individuals with highly sensitive skin were more likely than those with less sensitive skin to tan indoors in order to attain a pre-vacation tan [51]. There is also evidence from one study that indoor tanning is more common among people who have freckles and less common among women who reported having dark hair at the age of 20 years [51]. Indoor tanning has not been found to be associated with an individuals' eye color or the number of moles they have, although few studies have examined these factors [54, 61]. Thus, overall there is some evidence that indoor tanning varies according to individuals' physical characteristics such as their skin sensitivity and potentially also according to their hair color and presence of freckles.

3.4.3 *Behavioral Factors*

People who indoor tan may be more likely to engage in other sun exposure and health risk behaviors. There is evidence that individuals who tan indoors are also more likely to expose themselves to the sun by sunbathing or not wearing sun-protective clothing when outside on a sunny day [27, 37, 41, 46, 51, 58]. Similarly, they commonly report having more sunburns than individuals who do not tan indoors [37, 46, 51, 58]. However, the association between use of sunscreen and the prevalence of indoor tanning has been inconsistent across research studies and should be explored in future research [24, 25, 27, 37, 41, 51]. Several recent studies have found that indoor tanners are more likely to use sunless tanning products or procedures than individuals who do not tan indoors [36, 46, 48]. As discussed by Pagoto in Chap. 11 of this book, there is some evidence that individuals who tan outdoors can successfully adopt sunless tanning products as a substitute for sunbathing [68]. This provides optimism that indoor tanners could be encouraged to substitute indoor tanning with the use of sunless tanning, which would allow them to maintain a tan appearance and reduce exposure to health-damaging UV light.

Although there is relatively little relevant research, there is some indication that indoor tanners may be more likely to be physically active and to have a lower body mass index (BMI) than non-tanners [37, 46, 60, 61]. Many indoor tanners report tanning for appearance-related reasons—primarily in order to look attractive and healthy—which may in part explain their higher level of activity and lower BMI. Although results are not entirely consistent across all research studies, indoor tanners are also generally more likely to report a higher level of alcohol use, cigarette smoking, and use of other substances [21, 24, 37, 46, 47, 51, 56, 60, 61]. As noted in several recent studies, some indoor tanners meet criteria for tanning dependence, and these individuals may be at increased risk of substance use compared to indoor tanners who are not dependent on tanning [46, 47]. Greater understanding of the prevalence and psychosocial and biological mechanisms of indoor tanning dependence may inform the development of successful interventions to reduce indoor tanning behaviors.

3.4.4 *Social Factors*

There is consistent evidence that individuals are more likely to engage in indoor tanning if others in their immediate social environment—parents, siblings, friends, or merely acquaintances—also tan indoors [8, 21, 24, 25, 40, 48, 52, 56]. Social norms, cultural sanctions, and behavior-modeling tendencies (especially modeling after one's parents) may account for these findings. For some individuals, indoor tanning may be a social activity that they engage in with friends or family members. For further discussion of the role of social factors in indoor tanning, see Chap. 4 by Hillhouse and Turrisi in this book.

3.5 Measurement of Indoor Tanning

As shown in Table 3.1, studies of the prevalence of indoor tanning have varied widely in the time period of focus, including lifetime and past year prevalence, as well as current or typical use. The use of varying time periods and unqualified terms such as "current" or "regular" indoor tanning hinders comparison of prevalence estimates across studies [32, 52, 53]. Also, caution should be exercised when comparing the lifetime prevalence of indoor tanning across studies that focus on samples of varying ages or across age groups within a single study, as lifetime prevalence is inherently related to current age. Lifetime prevalence also does not provide an indication of the recency of indoor tanning. Thus, examination of potential age group differences in indoor tanning is best conducted using measures that inquire about frequency of tanning in a specific time period (e.g., in the past year) or that are conducted prospectively. There is a need for future prospective, longitudinal studies of indoor tanning to examine age-related changes in tanning.

Several studies have evaluated the test-retest reliability (consistency of responses over time) of indoor tanning survey items [69]. Overall, these studies have found acceptable levels of test-retest reliability (kappa reliability coefficients >0.70 out of 1.00) for indoor tanning lifetime use (ever vs. never) as well as frequency of use. Most studies used an interval between testings from 1 to 4 weeks, although one study used an interval from 1 to 3 years [70]. Using a sample of sorority and fraternity students who were surveyed during a spring semester, Dennis and colleagues examined the 2–4 week test-retest reliability of indoor tanning items that asked whether individuals had ever tanned indoors, as well as the number of indoor tanning sessions (with categories of 0, 1–5, 6–10, 11–15, 16–20, and >20 times) during each of the following time periods: before high school; during high school years; after age 18; during the past year; between New Year's and spring break; and total lifetime [69]. Kappa reliability coefficients varied from 0.63 for indoor tanning before high school to 0.97 for total lifetime use. Participants were first asked about indoor tanning during specific time periods in an effort to promote more accurate recall of total lifetime indoor tanning behaviors. However, it is unclear whether this increased the validity of reported lifetime indoor tanning, and future research is needed to test this notion experimentally. Test-retest reliability was also assessed for the typical clothing worn during indoor tanning (kappa = 0.72) and the use of facial protection (kappa = 0.85). Although there is good evidence for the acceptable test-retest reliability of self-reported indoor tanning behaviors, little research has examined the validity of such self-reports. Specifically, future research is warranted to examine the extent to which self-reported indoor tanning behaviors (e.g., frequency, duration, type of bed or booth used, and use of facial protection) correspond with objective behaviors.

Even when studies present data for the same time period they often use survey items with different wording, which may elicit different responses from study participants. In an effort to develop a uniform set of survey items that assesses a variety of skin cancer-related behaviors, including indoor tanning, the National Cancer Institute and Emory University cosponsored a 2005 meeting attended by researchers with expertise in skin cancer prevention. After compiling existing and newly-generated items, the researchers subjected all of the items to multiple rounds of cognitive interviewing with a sample of 31 individuals (87% female, aged 15–62 years, mean age = 25) who reported prior use of indoor tanning devices or sunless tanning products [71]. The study participants were confident in their ability to accurately recall the number of past year indoor tanning sessions, but were less confident about estimating their lifetime sessions (except if the number of lifetime sessions was very low). Based on the results of the cognitive interviewing, the research team generated a set of core indoor tanning survey items that they recommend for use in future research [71]. The items address lifetime use (ever vs. never) and frequency, past year frequency, and age at initiation of indoor tanning. The items can be adapted to focus on other time periods of interest. Additionally, the items may need to be adapted based on cultural or linguistic norms specific to populations of interest, such as using the term "solarium" when surveying individuals in Australia. Further, it is important to note that the items do not address other potentially important aspects of indoor tanning, including the duration of indoor tanning sessions, the type of bed or booth used, the use of facial protection, or the extent to which tanning sessions are condensed into a short period of time (e.g., in order to have a tan appearance for a specific event). Thus, the availability of core indoor tanning items such as those described by Lazovich and colleagues is an important first step in promoting the use of psychometrically sound measures that can be employed uniformly across studies conducted by different research groups [71]. However, the extent to which these items will be employed in research studies is unclear and, as previously noted, additional items are needed to address broader aspects of indoor tanning behaviors.

3.6 Key Points and Conclusions

Concurrent with the continued popularity of indoor tanning, a wealth of research has been conducted on the availability, prevalence, and correlates of indoor tanning in the past decade. For many individuals living in urban settings in the United States, an indoor tanning facility is more accessible to them than their favorite coffee shop or fast-food restaurant. Data on the prevalence and correlates of indoor tanning are available from studies conducted using a variety of samples in the Americas, multiple countries in Europe, and in Australia. It is not surprising that rates of indoor tanning vary widely across these diverse studies. For example, the prevalence of indoor tanning in the past year in population-based studies focusing primarily on adults has varied from 1% in Queensland (Australia) to 30% in Denmark. Among 11–18 year olds in the United States, rates of indoor tanning in the past year have varied from 11% to 26%, with rates among university students even higher at 33–60%. Identifying correlates of indoor tanning highlights groups that may benefit most from policy changes (such as the introduction of legislation restricting minors' access to indoor tanning) or targeted public health interventions. It also provides insight on intervention content and focus. In general, individuals who engage in indoor tanning are more likely than non-tanners to be female, aged 16–25 years, Caucasian, and to have low to moderate skin sensitivity to the sun. Other correlates of indoor tanning include knowing other people who tan indoors, and greater use of alcohol, cigarettes, and other substances. There are preliminary indications that correlates of indoor tanning may differ according to individuals' demographic characteristics. For example, as shown in Table 3.2, Heckman and colleagues found correlates of indoor tanning to vary according to the age group of focus [37]. Overall, they identified a greater number of correlates of indoor tanning among younger versus older adults, which is likely partly due to the lower tanning prevalence among older individuals. Several studies shown in Table 3.2 found certain correlates to be associated with indoor tanning only among men or women. However, there does not appear to be a systematic pattern with regard to such differences and additional research is warranted on this issue. Evidence that indoor tanning may be an addictive behavior for some individuals further suggests that interventions to reduce its prevalence will be maximally successful if they utilize a tailored approach that takes into account individuals' personal motivations for initiating and maintaining tanning. Further empirical research on correlates of indoor tanning among population subgroups will provide valuable information needed to develop interventions to reduce the prevalence of this health-damaging behavior.

Acknowledgments Thanks are due to Kristina Tatum and Marina Hanna for their valuable assistance in the preparation of this chapter. Elliot Coups is supported by career award 1K07CA133100 from the National Cancer Institute.

References

1. Albert MR, Ostheimer KG (2003) The evolution of current medical and popular attitudes toward ultraviolet light exposure: part 2. J Am Acad Dermatol 48(6):909–918
2. Albert MR, Ostheimer KG (2003) The evolution of current medical and popular attitudes toward ultraviolet light exposure: part 3. J Am Acad Dermatol 49(6):1096–1106
3. El Ghissassi F, Baan R, Straif K et al (2009) A review of human carcinogens—part D: radiation. Lancet Oncol 10(8):751–752
4. Swerdlow AJ, Weinstock MA (1998) Do tanning lamps cause melanoma? An epidemiologic assessment. J Am Acad Dermatol 38(1):89–98
5. International Agency for Research on Cancer (2007) The association of use of sunbeds with cutaneous malignant melanoma and other skin cancers: a systematic review. Int J Cancer 120(5):1116–1122

6. Karagas MR, Stannard VA, Mott LA et al (2002) Use of tanning devices and risk of basal cell and squamous cell skin cancers. J Natl Cancer Inst 94(3):224–226
7. US Food and Drug Administration (2010) Radiation-emitting products: sunlamps and sunlamp products (tanning beds/booths). http://www.fda.gov/Radiation-EmittingProducts/RadiationEmittingProductsandProcedures/HomeBusinessandEntertainment/ucm116447.htm. Accessed 4 May 2010
8. Cokkinides V, Weinstock M, Lazovich D et al (2009) Indoor tanning use among adolescents in the US, 1998 to 2004. Cancer 115(1):190–198
9. Knight JM, Kirinich AN, Farmer ER et al (2002) Awareness of the risks of tanning lamps does not influence behavior among college students. Arch Dermatol 138(10):1311–1315
10. Looking Fit (2010) Indoor tanning fact book 2009–2010. http://www.lookingfit.com/articles/looking-fit-fact-book-2009-2010.html. Accessed 4 May 2010
11. Palmer RC, Mayer JA, Woodruff SI et al (2002) Indoor tanning facility density in eighty U.S. cities. J Community Health 27(3):191–202
12. Hoerster KD, Garrow RL, Mayer JA et al (2009) Density of indoor tanning facilities in 116 large U.S. cities. Am J Prev Med 36(3):243–246
13. Jopson JA, Reeder AI (2008) An audit of Yellow Pages telephone directory listings of indoor tanning facilities and services in New Zealand, 1992–2006. Aust N Z J Public Health 32(4):372–377
14. Makin JK, Dobbinson SJ, Herd NL (2007) The increase in solariums in Australia, 1992–2006. Aust N Z J Public Health 31(2):191–192
15. Makin JK, Dobbinson SJ (2009) Changes in solarium numbers in Australia following negative media and legislation. Aust N Z J Public Health 33(5):491–494
16. Sinclair CA, Makin JK (2008) Sometimes it takes a loss of life to make a difference. BMJ 336(7635):73
17. Jalleh G, Donovan RJ, Lin C et al (2008) Changing perceptions of solaria and cancer risk: the role of the media. Med J Aust 188(12):735
18. Schneider S, Kramer H (2010) Who uses sunbeds? A systematic literature review of risk groups in developed countries. J Eur Acad Dermatol Venereol 24(6):639–648
19. Bandi P, Cokkinides VE, Weinstock MA et al (2010) Sunburns, sun protection and indoor tanning behaviors, and attitudes regarding sun protection benefits and tan appeal among parents of U.S. adolescents-1998 compared to 2004. Pediatr Dermatol 27(1):9–18
20. Robinson JK, Kim J, Rosenbaum S et al (2008) Indoor tanning knowledge, attitudes, and behavior among young adults from 1988–2007. Arch Dermatol 144(4):484–488
21. Bagdasarov Z, Banerjee S, Greene K et al (2008) Indoor tanning and problem behavior. J Am Coll Health 56(5):555–561
22. Clough-Gorr KM, Titus-Ernstoff L, Perry AE et al (2008) Exposure to sunlamps, tanning beds, and melanoma risk. Cancer Causes Control 19(7):659–669
23. Lucci A, Citro HW, Wilson L (2001) Assessment of knowledge of melanoma risk factors, prevention, and detection principles in Texas teenagers. J Surg Res 97(2):179–183
24. Lazovich D, Forster J, Sorensen G et al (2004) Characteristics associated with use or intention to use indoor tanning among adolescents. Arch Pediatr Adolesc Med 158(9):918–924
25. Stryker JE, Lazovich D, Forster JL et al (2004) Maternal/female caregiver influences on adolescent indoor tanning. J Adolesc Health 35(6):528 e521–529
26. Zeller S, Lazovich D, Forster J et al (2006) Do adolescent indoor tanners exhibit dependency? J Am Acad Dermatol 54(4):589–596
27. Ramirez MA, Warthan MM, Uchida T et al (2003) Double exposure: natural and artificial ultraviolet radiation exposure in beachgoers. South Med J 96(7):652–655
28. Lazovich D, Sweeney C, Forster J (2005) Prevalence of indoor tanning use in Minnesota, 2002. Arch Dermatol 141(4):523–524
29. Benvenuto-Andrade C, Zen B, Fonseca G et al (2005) Sun exposure and sun protection habits among high-school adolescents in Porto Alegre. Brazil Photochem Photobiol 81(3):630–635
30. Buchanan N, Leisenring W, Mitby PA et al (2009) Behaviors associated with ultraviolet radiation exposure in a cohort of adult survivors of childhood and adolescent cancer: a report from the Childhood Cancer Survivor Study. Cancer 115(18 Suppl):4374–4384

31. Danoff-Burg S, Mosher CE (2006) Predictors of tanning salon use: behavioral alternatives for enhancing appearance, relaxing and socializing. J Health Psychol 11(3):511–518
32. Hillhouse J, Stapleton J, Turrisi R (2005) Association of frequent indoor UV tanning with seasonal affective disorder. Arch Dermatol 141(11):1465
33. Sheehan DJ, Lesher JL Jr (2005) The effect of sunless tanning on behavior in the sun: a pilot study. South Med J 98(12):1192–1195
34. Ting W, Schultz K, Cac NN et al (2007) Tanning bed exposure increases the risk of malignant melanoma. Int J Dermatol 46(12):1253–1257
35. Johnson KR, Heilig LF, Hester EJ et al (2006) Indoor tanning attitudes and practices of US dermatologists compared with other medical specialists. Arch Dermatol 142(4):465–470
36. Brooks K, Brooks D, Dajani Z et al (2006) Use of artificial tanning products among young adults. J Am Acad Dermatol 54(6):1060–1066
37. Heckman CJ, Coups EJ, Manne SL (2008) Prevalence and correlates of indoor tanning among US adults. J Am Acad Dermatol 58(5):769–780
38. Coups EJ, Manne SL, Heckman CJ (2008) Multiple skin cancer risk behaviors in the U.S. population. Am J Prev Med 34(2):87–93
39. Stapleton J, Turrisi R, Hillhouse J (2008) Peer crowd identification and indoor artificial UV tanning behavioral tendencies. J Health Psychol 13(7):940–945
40. Hoerster KD, Mayer JA, Woodruff SI et al (2007) The influence of parents and peers on adolescent indoor tanning behavior: findings from a multi-city sample. J Am Acad Dermatol 57(6):990–997
41. Stryker JE, Yaroch AL, Moser RP et al (2007) Prevalence of sunless tanning product use and related behaviors among adults in the United States: results from a national survey. J Am Acad Dermatol 56(3):387–390
42. Cafri G, Thompson JK, Jacobsen PB et al (2009) Investigating the role of appearance-based factors in predicting sunbathing and tanning salon use. J Behav Med 32:532–544
43. Poorsattar SP, Hornung RL (2007) UV light abuse and high-risk tanning behavior among undergraduate college students. J Am Acad Dermatol 56(3):375–379
44. Ma F, Collado-Mesa F, Hu S et al (2007) Skin cancer awareness and sun protection behaviors in white Hispanic and white non-Hispanic high school students in Miami. Florida Arch Dermatol 143(8):983–988
45. Hillhouse J, Turrisi R, Shields AL (2007) Patterns of indoor tanning use: implications for clinical interventions. Arch Dermatol 143(12):1530–1535
46. Heckman CJ, Egleston BL, Wilson DB et al (2008) A preliminary investigation of the predictors of tanning dependence. Am J Health Behav 32(5):451–464
47. Mosher CE, Danoff-Burg S (2010) Addiction to indoor tanning: relation to anxiety, depression, and substance use. Arch Dermatol 146(4):412–417
48. Gordon D, Guenther L (2009) Tanning behavior of London-area youth. J Cutan Med Surg 13(1):22–32
49. Castilho IG, Sousa MA, Leite RM (2010) Photoexposure and risk factors for skin cancer: an evaluation of behaviors and knowledge among university students. An Bras Dermatol 85(2):173–178
50. Bataille V, Boniol M, De Vries E et al (2005) A multicentre epidemiological study on sunbed use and cutaneous melanoma in Europe. Eur J Cancer 41(14):2141–2149
51. Ezzedine K, Malvy D, Mauger E et al (2008) Artificial and natural ultraviolet radiation exposure: beliefs and behaviour of 7200 French adults. J Eur Acad Dermatol Venereol 22(2):186–194
52. Bränström R, Ullen H, Brandberg Y (2004) Attitudes, subjective norms and perception of behavioural control as predictors of sun-related behaviour in Swedish adults. Prev Med 39(5):992–999
53. Devos SA, Baeyens K, Van Hecke L (2003) Sunscreen use and skin protection behavior on the Belgian beach. Int J Dermatol 42(5):352–356
54. Dissel M, Rotterdam S, Altmeyer P et al (2009) Indoor tanning in North Rhine-Westphalia Germany: a self-reported survey. Photodermatol Photoimmunol Photomed 25(2):94–100

55. Hamlet N, Kennedy K (2004) Reconnaissance study of sunbed use by primary school children in Lanarkshire. J Public Health (Oxford) 26(1):31–33
56. Mackay H, Lowe D, Edwards DJ et al (2007) A survey of 14 to 16 year olds as to their attitude toward and use of sunbeds. Health Educ J 66:141–152
57. Börner FU, Schutz H, Wiedemann P (2009) A population-based survey on tanning bed use in Germany. BMC Dermatol 9:6
58. Køster B, Thorgaard C, Clemmensen IH et al (2009) Sunbed use in the Danish population in 2007: a cross-sectional study. Prev Med 48(3):288–290
59. Roberts DL, Foley K (2009) Sunbed use in children-time for new legislation? Br J Dermatol 161(1):193–194
60. Diehl K, Litaker DG, Greinert R et al (2010) The prevalence of current sunbed use and user characteristics: the SUN-study 2008. Int J Public Health 55(5):513–516
61. Schneider S, Zimmerman S, Diehl K et al (2009) Sunbed use in German adults: risk awareness does not correlate with behaviour. Acta Derm Venereol 89(5):470–475
62. Thomson CS, Woolnough S, Wickenden M et al (2010) Sunbed use in children aged 11–17 in England: face to face quota sampling surveys in the National Prevalence Study and Six Cities Study. BMJ 340:c877
63. Paul CL, Girgis A, Tzelepis F et al (2004) Solaria use by minors in Australia: is there a cause for concern? Aust N Z J Public Health 28(1):90
64. Dobbinson S, Wakefield M, Hill D et al (2008) Prevalence and determinants of Australian adolescents' and adults' weekend sun protection and sunburn, summer 2003–2004. J Am Acad Dermatol 59(4):602–614
65. Lawler SP, Kvaskoff M, DiSipio T et al (2006) Solaria use in Queensland, Australia. Aust N Z J Public Health 30(5):479–482
66. Zahnd WE, Goldfarb J, Scaife SL et al (2010) Rural-urban differences in behaviors to prevent skin cancer: an analysis of the Health Information National Trends Survey. J Am Acad Dermatol 62(6):950–956
67. National Conference of State Legislatures (2010) Tanning restrictions for minors. A state-by-state comparison. http://www.ncsl.org/programs/health/tanningrestrictions.htm. Accessed 4 May 2010
68. Pagoto SL, Schneider KL, Oleski J et al (2010) The sunless study: a beach randomized trial of a skin cancer prevention intervention promoting sunless tanning. Arch Dermatol 146(9):979–984
69. Dennis LK, Kim Y, Lowe JB (2008) Consistency of reported tanning behaviors and sunburn history among sorority and fraternity students. Photodermatol Photoimmunol Photomed 24(4):191–198
70. Westerdahl J, Anderson H, Olsson H et al (1996) Reproducibility of a self-administered questionnaire for assessment of melanoma risk. Int J Epidemiol 25(2):245–251
71. Lazovich D, Stryker JE, Mayer JA et al (2008) Measuring nonsolar tanning behavior: indoor and sunless tanning. Arch Dermatol 144(2):225–230

Chapter 4
Motivations for Indoor Tanning: Theoretical Models

Joel Hillhouse and Rob Turrisi

Abstract This chapter reviews the literature applying health behavior theories to indoor tanning. Few studies have tried to fit full versions of health behavior models to indoor tanning. Theoretical models from the family of theories referred to as the reasoned action approach (e.g., theory of planned behavior, behavioral alternative model, prototype willingness model, etc.) have been most commonly used to study indoor tanning. Results indicate that these models fit indoor tanning data moderately to extremely well. Two lesser known models, problem behavior theory and the terror management health model, have also demonstrated a reasonable fit. Two other common models, the health belief model and social cognitive theory, have never been fully tested with indoor tanning. However, key constructs from these models (e.g., perceived susceptibility and threat, modeling) have been used to understand indoor tanning. Empirical research conducted represents a solid start toward developing strong, comprehensive models of indoor tanning that can guide intervention efforts. This initial work needs to be expanded by conducting longitudinal studies and by including a broader age range in studies because the majority of existing work has focused on young adults. Incorporating findings related to tanning dependency, peer group affiliation, media influences and other constructs into these foundational models will also improve our understanding and ability to develop efficacious interventions to reduce engagement in this health risk behavior.

J. Hillhouse (✉)
Department of Community and Behavioral Health, East Tennessee State University,
Box 70674, Johnson City, TN 37614, USA
e-mail: hillhous@etsu.edu

R. Turrisi
Biobehavioral Health & Prevention Research Center, The Pennsylvania State University,
109 S. Henderson, University Park, PA 16801, USA
e-mail: rturrisi@psu.edu

C.J. Heckman and S.L. Manne (eds.), *Shedding Light on Indoor Tanning*,
DOI 10.1007/978-94-007-2048-0_4, © Springer Science+Business Media B.V. 2012

Keywords Indoor tanning • Health behavior • Theoretical models • Health belief model • Problem behavior theory • Reasoned action approach • Theory of planned behavior • Prototype theory • Terror management • Social cognitive model • Behavioral alternative model • Hillhouse • Turrisi • Integrative model

Abbreviations

TPB	Theory of planned behavior
TRA	Theory of reasoned action
PBC	Perceived behavioral control
BAM	Behavioral alternatives model
PWM	Prototype willingness model
PBT	Problem behavior theory
TMT	Terror management health model
RCT	Randomized controlled trial
UV	Ultraviolet

4.1 Introduction

Research focusing on indoor tanning has increased dramatically over the last 20 years. To illustrate this point, a review of the literature revealed one article published in 1990 dealing with indoor tanning [1]. However, in 2009, there were 24 articles on this topic. The increasing popularity of indoor tanning together with its recent reclassification as a Group 1 carcinogen (equivalent to smoking, arsenic and mustard gas) by the International Agency for Research on Cancer is likely to lead to further growth in the empirical study of this behavioral risk factor for skin cancer [2]. As the study of indoor tanning moves forward, it is critical that the research be accompanied by the use of strong theoretical models supported by solid empirical data [3]. The purpose of this chapter is to review the use of theoretical models applied to the understanding of indoor tanning and interventions directed at reducing this behavior.

The most common reason cited for indoor tanning is the belief that tans make one more attractive [4–12]. Relaxation/stress relief aspects of the indoor tanning experience are the second most cited motive for indoor tanning [5, 6, 13–18], with other commonly studied predictors including parental attitudes toward indoor tanning [19], the idea that a tan is healthy-looking [4, 12], and perceived advantages of indoor tanning over tanning in the sun [4, 16].

While the study of isolated constructs associated with indoor tanning is important in model-building, the Society of Prevention Research requires prevention interventions to be grounded in strong, empirically-tested theoretical models of health behavior as one of its standards of evidence [20]. Only a handful of indoor tanning studies have been published to date that have attempted to fit full versions (i.e., all of the constructs included in the theory) of empirically-supported models to indoor tanning behaviors. These will be reviewed first, and then we will summarize the findings from studies that have explored important, but isolated, constructs from established theoretical models.

4.2 Theories and Models

4.2.1 Theory of Planned Behavior

The theory of planned behavior (TPB; [21]) represents an extension of the theory of reasoned action (TRA; [22]), which hypothesizes that behavior is best predicted by corresponding behavioral intentions. Intentions, in turn, are influenced by attitudes toward the behavior, and subjective norms concerning the behavior (i.e., perceived attitudes of significant others toward the behavior weighted by the individual's motivation to comply with these attitudes). The TPB adds the construct of perceived behavioral control (PBC), derived from social cognitive theory's self-efficacy construct. PBC represents an individual's perceived ease or difficulty in successfully performing the behavior in the future.

Hillhouse and colleagues utilized a longitudinal design to assess the degree of fit of the TPB to several UV-risk behaviors, including indoor tanning, in a group of female and male college students [23]. Participants were initially assessed regarding their indoor tanning attitudes and subjective norms. One week later they were assessed on their perceived control and intentions toward indoor tanning, followed one week later by assessments of actual indoor tanning use. Indoor tanning intentions were found to significantly predict subsequent behavior ($R^2=.71$). Significant unique effects were observed for attitudes ($b=.08$, $t=4.64$, $p<.01$, semipart $r^2=.07$) and perceived behavioral control ($b=.55$, $t=6.28$, $p<.01$, semipart $r^2=.13$) in the prediction of indoor tanning intentions. Together, the variables accounted for 63% of the indoor tanning intention variance. Further examination revealed a significant interaction between attitudes and perceived behavioral control. As PBC increased, the relationship between attitudes and intentions strengthened, indicating that feelings toward indoor tanning influence intentions most when the individual believes they have control over the behavior. Similarly, the relationship between subjective norms and intentions was moderated by PBC. That is, as PBC increased, the relationship between subjective norms and intentions increased. These results suggest that perceptions of others' beliefs about indoor tanning were influential when people believed they were able to control their own indoor tanning behavior.

This study was replicated in another sample of college students, with the added constructs of appearance motivation, self-monitoring, and health orientation [24]. As predicted by the TPB, greater intentions significantly predicted more indoor tanning behavior ($R^2=.55$), and attitudes toward indoor tanning, subject norms, and perceived behavioral control significantly positively predicted intentions to indoor tan ($R^2=.58$).

Appearance motivation was a stronger positive predictor of indoor tanning attitudes than was health orientation. Self-monitoring interacted with subjective norms in the prediction of indoor tanning intentions with high self-monitors (i.e., individuals primarily guided by external cues) demonstrating stronger norm-intention relationships than low self-monitors (i.e., individuals guided by internal cues). Given that high self-monitors are more likely to listen to the opinions of significant others in their lives [25], it is not surprising that subjective norms would more strongly affect indoor tanning intentions in high self-monitors.

4.2.2 Behavioral Alternatives Model

The behavioral alternatives model (BAM; [26]) has been used to model indoor tanning behavior in two studies [27, 28] and has been the basis for two behavioral interventions focused on reducing indoor tanning behavior in young people [29, 30]. The BAM is based on the fact that in any given situation individuals have available several behavioral options. For example, a young person interested in improving her appearance for the prom has the option to indoor tan but also the options to use spray-on tanning, or to choose clothing that does not require a complimentary tan. The BAM theorizes that individuals will choose those behaviors from amongst these alternatives that are most appealing to them. Understanding indoor tanning decisions involves modeling indoor tanning behavior together with the other viable behavioral options being considered (such as spray tanning). Interventions based on the BAM seek to reduce the appeal of risky behaviors such as indoor tanning while simultaneously increasing the appeal of viable alternative behaviors (See Chap. 9 by Turrisi and colleagues). Behavior change is predicted to occur when the healthy alternative behaviors are perceived as more appealing than the risky behavior.

Hillhouse and colleagues used constructs from the TPB (i.e., behavioral attitudes and subjective norms) to model indoor tanning alternatives for enhancing appearance (i.e., indoor tanning, fashionable clothing, exercise, grooming, weight control) in a group of college students [28]. Results indicated that individuals who feel favorable toward indoor tanning but unfavorable toward wearing fashionable clothing as a good way to enhance appearance were more likely to indoor tan ($F\,(9,221) = 12.85$, $p < .001$, $R^2 = .34$). Further analysis found that those higher in appearance motivation, with greater perceptions that their peers were indoor tanning, and lower in perceptions of the risk of skin appearance damage from tanning were significantly more likely to have positive attitudes toward indoor tanning. Fashionable clothing-use attitudes were predicted by appearance motivation, perceiving advantages in using clothing to enhance appearance (e.g., "People notice me more when I wear attractive clothing") and perceiving fewer disadvantages to using clothing to improve appearance (e.g., "Buying nice clothing is expensive").

Danoff-Burg and colleagues also used the BAM to investigate cognitive predictors of indoor tanning in college students by exploring the association of indoor tanning during the past year with attitudes toward alternatives for enhancing physical appearance (i.e. using tanning salons, self-tanning products, clothing, diet, exercise), alternatives for relaxing (i.e. going to tanning salons; watching television or movies; practicing meditation, yoga or deep breathing; going out with friends; engaging in a favorite hobby), and attitudes toward alternatives for socializing (i.e. going to tanning salons, going to a gym, going to restaurants or bars, going to parties) [27]. Unlike the earlier study [28], they did not find that the non-indoor tanning appearance alternatives significantly predicted indoor tanning use. Instead, they report that favorable attitudes toward engaging in a hobby to relax ($b=-0.21$, $t=-2.89$, $p<.01$) and going to the gym to socialize ($b=-0.17$, $t=-2.12$, $p<.01$) were negatively associated with indoor tanning frequency.

4.2.3 Prototype Willingness Model

The prototype willingness model (PWM) developed by Gibbons and Gerrard hypothesizes that young people with more positive prototypes of a particular behavior are more willing to engage in the behavior themselves when the opportunity arises [31]. Prototypes represent social images individuals have about the type of person who would typically engage in a particular behavior. For example, an individual who believes the typical indoor tanner is attractive and cool would be expected to be more willing to take advantage of indoor tanning opportunities than someone who considers the representative indoor tanner as immature and careless. The PWM utilizes the construct of willingness to capture those situations where young people may engage in a behavior that they did not originally intend. An example of greater willingness would be a sorority member who agrees to go along with her sorority sisters to a tanning salon even though she was not intending to indoor tan before that time.

Testing this model, Gibbons and colleagues used UV photography to highlight UV exposure-caused skin damage in participants [32]. The authors examined whether a tanning cognition index of health risk composed of variables suggested by the PWM (i.e., tanning attitudes, tanner prototypes, and willingness to engage in risky UV exposure) mediated the effects of the UV photograph intervention. Their results indicated that this tanning cognition index significantly mediated the effect of intervention condition on indoor tanning use 4 weeks later. Specifically, the intervention reduced favorable attitudes toward tanning, increased perceived vulnerability to appearance-related skin damage, produced less favorable tanning prototypes, and led to marginally less willingness for UV exposure immediately following the intervention. Four weeks later perceived vulnerability to appearance-related skin damage was still increased relative to controls. Changes in UV-exposure willingness had become a significant mediator at follow-up but attitude or prototype mediation was no longer significant. The fact that the intervention effects were mediated by some of the PWM variables in the expected direction provide partial support for the validity of this model for indoor tanning behavior. However, future empirical research needs to be designed to fully test this model with indoor tanning.

4.2.4 *Problem Behavior Theory*

Problem behavior theory (PBT; [33]) hypothesizes that risk behaviors are due to interactions in the domains of personality (e.g., self-esteem, sensation-seeking, etc.), environment (e.g., peer-approval, norms, parental control, etc.), and behavior (e.g., smoking, drug use, violence, etc.). PBT proposes that elements of each of these major domains either encourage or discourage risky behaviors, resulting in a risk level of engaging in the behavior. One recent cross-sectional study in 743 young adults found evidence that elements from each of these domains were predictive of past indoor tanning and future intentions [34]. Specifically, the authors found positive relationships in the personality domain between sensation seeking and tanning image beliefs (i.e., tanned look, such as wanting to look healthier, feel more confident, and appear attractive) with past indoor tanning (IT) behavior and future IT intentions. The authors found significant relationships in the environmental domain (i.e., acquaintance and friend use of indoor tanning) with tanning bed use and intentions. Greater use of tanning beds by friends and acquaintance was correlated with past tanning bed use and future intentions to use a tanning bed. Lastly, they report associations in the behavior domain, in that greater alcohol use was positively related to past indoor tanning behavior and that both smoking and drinking behavior were positively associated with indoor tanning intentions.

4.2.5 *Terror Management Health Model*

The terror management health model (TMT; [35]) is based on the idea that individuals strive to enhance their self-esteem through endeavors such as personal accomplishments and appearance enhancements as a means of decreasing anxiety that arises from an awareness of personal mortality. Tanning, in this conceptualization, is conceived of as one possible means to increase self-esteem through its enhancement of appearance. Individuals with increased awareness of personal mortality are hypothesized to be more strongly motivated to improve their self-esteem. Therefore, in contexts where tans are portrayed as attractive, tanning intentions are expected to rise among those with higher anxiety related to their fear of death. TMT proposes that health-focused messages may inadvertently increase tanning motivations by increasing the salience of personal mortality. The studies utilizing TMT for understanding tanning intentions manipulate mortality awareness by having the participants respond in writing to questions about their eventual death (e.g., "Please briefly describe the emotions that the thought of your own death arouses in you") [36, 37]. In both these studies, some participants received materials that emphasized the attractiveness of tanned skin, while others were exposed to materials that stressed the appearance advantages of pale skin. In support of the TMT, both studies reported that participants with higher death-related anxiety reported higher tanning intentions when given material that emphasized positive aspects of tanned skin but lower intentions when they received information that emphasized the attractiveness of pale skin.

The model appears counter-intuitive to traditional health-based approaches to skin cancer prevention in that individuals are expected to tan more to reduce death anxiety despite the fact that tanning increases actual mortality risk. However, the model is based on the idea that individuals do not always rely on rational decision-making processes when making behavioral choices.

4.2.6 Symbolic Interactionism

Using symbolic interactionism [38], Vannini and McCright, explored the meaning of tanning and indoor tanning through qualitative methods that included in-depth interviews with college male and female tanners [39]. Symbolic interactionism emphasizes the symbolic meanings of objects and behaviors in the context of specific social interactions. In this article, the authors explored the "medical frame" (i.e., interpretation and communication of indoor tanning according to principals of medicine, health, and wellness) versus the "seductive frame" (i.e., interpretation and communication of tanning in terms of its physical attractiveness, athletic fitness, self-esteem, and self-confidence). Results indicated that the medical frame was not as powerful as the seduction frame for explaining tanning behavior and the tanned body. Indoor tanners reported that they were often told they look "good", "sexy", "beautiful", "young", "healthy" and that these comments were important for a positive self-concept. The seductive frame interpretation of tanning was more compelling to the tanners than the medical frame. The interviews also revealed that the tanned look had value for those around the tanner who like to look at "beautiful, young, healthy, and fit others." Indoor tanners also reported deriving pleasure from "seducing" others in this way.

4.2.7 Objectification Theory

The idea that sexual objectification of women driven by cultural standards and the media lead to women viewing themselves and their bodies as objects to be examined and watched is termed objectification theory [40]. Stapleton and colleagues used this model to evaluate indoor tanning behavior in a cross-sectional study of 155 college females [41]. They used the objectified body consciousness scale [42] to measure self-objectification (i.e., treating oneself as an object of desire) and body shame. Self-objectification was measured through ratings of statements assessing how concerned the person is with how she looks to other people. Body shame was assessed with items measuring feelings of shame associated with activities and situations such as not exercising and not looking good. Using structural equation modeling, the authors tested the relationship between body objectification constructs and indoor tanning. Results indicated a good model fit (comparative fit index). The self-objectification variable was significantly related to higher bodily shame, and bodily shame was significantly related to intentions to indoor tan, which were related to indoor tanning behavior.

4.2.8 Constructs from Health Models

The health belief model (HBM; [43]) and social cognitive theory (SCT; [44]) represent two other important theoretical models that have been widely used to study a variety of health behaviors. The HBM is based on the idea that individuals will engage in health behaviors that they believe will provide benefits in relation to health threats they perceive as serious and to which they believe they are vulnerable. Key components of this model include perceived susceptibility and perceived severity of health threats, perceived benefits versus barriers of health actions, cues that signal individuals to initiate health actions and self-efficacy or beliefs about the individual's ability to successfully complete specific health actions. SCT focuses on how we learn from models (e.g., significant others, peers, and media) through observing their behaviors. SCT also describes the influence of expectations, which are defined as an individual's anticipated short- and long-term outcomes from engaging in a behavior, and expectancies, the values the individual places on these outcomes.

We found no studies that examined full versions of either the HBM or SCT in relation to indoor tanning behavior. However, there are a number of articles that utilize some key constructs from each of these theories to explore indoor tanning. Specifically, the constructs of perceived susceptibility and threat from the HBM and parental and peer modeling from the SCT have been explored in a number of indoor tanning articles.

Several studies have explored the relationship of perceived susceptibility/threat to skin cancer and/or skin damage to indoor tanning behavior. The majority of these studies focused on young adult women. Two of these examined a variable they labeled skin harm which represented a combination of items assessing the relationship of indoor tanning to skin cancer development and appearance-related skin damage. Greene and Brinn reported that both perceived susceptibility and threat of skin harm were significant predictors of indoor tanning intentions [45]. Women reporting greater perceived threat were less likely to report intending to indoor tan. However, those reporting greater susceptibility to skin harm reported greater intentions to indoor tan. While this latter finding seems counter-intuitive, it likely reflects the fact that individuals who intend to indoor tan are more aware of the connection between tanning and skin harm than their peers who may have little interest in the behavior. Hillhouse and colleagues found that the perceived susceptibility and severity of skin harm did not significantly predict indoor tanning behavior [16].

Two other studies examined the relationship of perceived susceptibility to cancer with indoor tanning. Coups and colleagues [46] and Heckman and colleagues [47] both reported that young adults who perceived themselves at greater risk for cancer also reported more skin cancer risk behaviors including indoor tanning.

However, Cafri and colleagues found that perceiving indoor tanning as a threat for skin aging was significantly negatively related to intentions to indoor tan [8]. These results likely differ due to the differences in focus on skin cancer risk versus skin aging. One explanation is that appearance damage is more salient and threatening to young adults than cancer risk which is perceived as a distant threat. This is supported by the good success in reducing tanning intentions and behaviors of appearance-focused interventions reported in the literature [29, 30, 32]. We found no studies that explored perceived benefits versus barriers or cues to action in relation to indoor tanning.

4.2.9 Social Influence Models

Numerous studies have reported relationships between indoor tanning behavior in individuals and the tanning behavior of significant role models including parents and friends. For example, Cokkinides and colleagues have reported that there is a positive relationship between adolescents' indoor tanning behavior and their parents' use of indoor tanning facilities [48, 49]. Stryker and colleagues report a similar relationship between the maternal/female caregiver and both female and male adolescent tanning behavior [50]. Hoerster and colleagues also report an association between parents' indoor tanning behavior and the behavior of teenage offspring [19]. In a recent study from our laboratory, we found that adolescent women who had initiated indoor tanning accompanied by their mothers were more likely to become frequent, habitual indoor tanners by young adulthood [51]. These findings further support the potential importance of parental modeling, particularly between mothers and teen offspring.

Positive relationships with peer tanning have also been reported. Hoerster and colleagues reported a strong relationship between individuals' reports of their friends' and their own tanning behavior [19]. Similar relationships between adolescents' perceptions of tanning by their peers and their own tanning practices were found in a study by Lazovich and colleagues [52].

4.2.10 Hybrid Approaches

In contrast to using one theory or model to guide their research, other researchers have attempted to utilize constructs from a variety of different models. For example, Cafri and colleagues [53] took another approach to modeling indoor tanning behavior by attempting to incorporate multiple key constructs from several theories into a single model. They used constructs from the theory of reasoned action, variables derived from the literature examining body image (e.g., beliefs about thinness and attractiveness), the health belief model, and protection motivation theory to explore indoor tanning in a cross-sectional study of 589 college women. Using structural

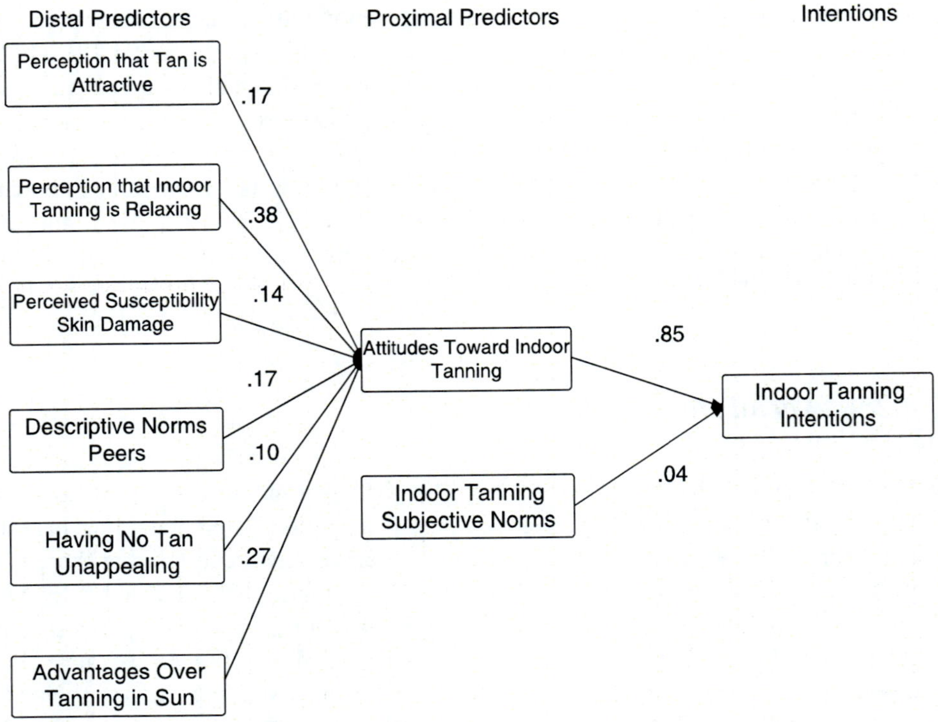

Fig. 4.1 Model of indoor tanning [54]

equation modeling, the authors found that indoor tanning attitudes predicted indoor tanning intentions, which in turn predicted indoor tanning behavior, supporting the theory of reasoned action. However, congruent with the work found in the body image literature, the results also indicated that beliefs that tans are attractive, tanning reduces the effects of acne, and a tan makes one look thinner mediated a relationship between sociocultural influences on tanning (i.e., perceptions of celebrity and peer tanning and of family and significant others' attitudes towards tanning) and indoor tanning behavior. Lastly, concerns with skin aging and appearance-related skin damage and indoor tanning intentions were found to mediate the influence of perceived threats of developing skin cancer from tanning on indoor tanning behavior, supporting both the health belief model and protection motivation theory. The authors did not explicitly choose to test one theory versus another, instead choosing to combine the constructs from a number of theoretical models. Future studies may want to more directly compare these models.

Our own laboratory has also recently completed work on a study [54] that examined the fit of variables from several models to indoor tanning behavior in a cross-sectional study of 384 young adults (see Fig. 4.1). Focus groups and preliminary empirical studies were used to delineate key indoor tanning variables to be included in the model. Consistent with the TPB, the model in Fig. 4.1 predicts that indoor tanning attitudes and subjective norms relate to indoor tanning intentions. Also consistent with the constructs from the TPB, the hybrid model explores perceived advantages and disadvantages of indoor tanning (e.g., beliefs that tans are attractive, whether indoor tanning is relaxing and stress relieving, whether indoor tanning can increase risk for cancer, whether indoor tanning can harm skin appearance, if friends and celebrities indoor tan, whether not having a tan is unappealing) as distal predictors of indoor tanning attitudes. However, the model also includes constructs from the behavioral alternative model (e.g., whether indoor tanning has advantages over tanning in the sun). The model was explored using structural equation modeling. The final model demonstrated a good fit to the data (e.g., CFI = .996, non-significant χ^2) and accounted for over 70% of the indoor tanning intention variance. Indoor tanning attitudes, the most important proximal predictor of intentions, were significantly related to beliefs that tans are attractive (r = .17), perceptions that indoor tanning is relaxing (r = .38), perceived susceptibility to appearance damage to the skin (r = −.14), indoor tanning descriptive norms (r = .17), beliefs that not being tan is unappealing (r = .10), and beliefs that indoor tanning has advantages over tanning in the sun (r = .27). Overall, these beliefs were able to account for 62% of the indoor tanning attitude variance (see Fig. 4.1). However it is important to note that in this study other theoretical constructs, models, and possible relations between variables were not explored (e.g., perceived behavioral control, whether there were direct effects of the constructs conceptualized as distant predictors on indoor tanning intentions). This was done in part to stay as close to the original theories (TPB and BAM) to guide the relationships between variables as possible.

Together, these two studies show some promise in combining constructs from different theories in an attempt to provide optimal explanation of indoor tanning tendencies.

4.3 Summary

Applying full versions of empirically validated theoretical models is still relatively uncommon in the study of indoor tanning behavior. Therefore, it is not yet possible to definitively conclude which models most effectively explain indoor tanning. Examining the available evidence, the TPB demonstrates a strong and consistent fit, at least in young adult populations. The two studies utilizing the TPB were able to account for approximately 60% of the indoor tanning intention variance and 50–70% of indoor tanning behavioral variance [23, 24]. Indoor tanning attitudes were the strongest predictors of intentions in both TPB studies. It is important to note that these studies examined attitudes toward indoor tanning specifically as opposed to measuring attitudes toward tanning in general or the perception that tans are attractive as has been done in numerous other articles in the literature [5, 9, 11, 12, 48, 49, 52, 55]. This is significant given that beliefs about tans and attitudes toward tanning in general are not identical to attitudes toward indoor tanning. This is because there are a variety of ways to obtain a tan including sunless tanning, sunbathing and of course indoor tanning. Understanding the specific motivations of indoor tanners is important for designing prevention interventions targeting this special subgroup of tanners. For example, it is possible for someone to have a negative attitude toward tanning outdoors but a positive attitude toward indoor tanning due to the misinformation disseminated by the tanning industry that indoor tanning is safe and even healthful. It will also be important to consider attitudes toward tans separately from indoor tanning when considering offering sunless tanning products as a healthier alternative.

The integrative model, which incorporates the core TPB constructs, has the advantage of pinpointing specific beliefs and perceptions that underlie tanning motivations. Compared to more general beliefs, these beliefs are more easily influenced in short-term educational interventions. Therefore, the IM can provide specific guidance in the development of prevention interventions. The one study that specifically tested this model found that it fit the available data quite well, accounting for approximately 70% of the indoor tanning intention variance [54]. Variables underlying indoor tanning decisions identified in this study included beliefs that tans are attractive, indoor tanning is relaxing, indoor tanning has advantages (e.g., is more healthy and convenient) over sunbathing, one's peers are tanning, being pale is unattractive, and indoor tanning potentially can lead to appearance-related skin damage.

The behavioral alternatives model, a variation on the TPB that allows the consideration of healthy alternatives, has also fared well in the couple of empirical studies conducted thus far. Both Hillhouse and colleagues [28] and Danoff-Burg and Mosher [27] found support for its utility in understanding indoor tanning behavior in young adults. Interestingly, Danoff-Burg and Mosher were unable to replicate the earlier findings that appearance-related alternatives were important for predicting indoor tanning. Instead, they found that social and stress-relief alternatives were significantly negatively related to indoor tanning behavior. The earlier Hillhouse study [28] did not assess social and hobby alternatives. If they had, it is possible that such non-appearance alternatives would have proved significant, reducing the effects of the appearance alternatives on indoor tanning motivations. Further study is needed to clarify this discrepancy.

One advantage of the BAM is that it provides empirical data on behaviors that may serve as healthy substitutes for indoor tanning. It is unlikely that cultural or individual beliefs about the importance of attractiveness for young people will be changed in short-term interventions. When interventions do not provide healthy alternative means to improve attractiveness, many who give up tanning in response to interventions will likely return to tanning as they continue to seek to enhance their appearance. Providing healthy alternative means to achieve this goal is more likely to lead to long-term changes in behavior.

Using a model combining the IM and BAM, an appearance-focused intervention has been developed that has been able to reduce indoor tanning by 35–50% in young adults [29, 30]. Details on this intervention are covered more thoroughly in another chapter in this volume (Turrisi et al., Chap. 9). The beliefs underlying indoor tanning attitudes identified from the IM (i.e., beliefs that tans are attractive, indoor tanning is relaxing, one's peers are tanning, and indoor tanning potentially can lead to appearance-related skin damage) have been shown to mediate the effects of the intervention on long-term indoor tanning intentions and behavior in randomized controlled trials. These studies have also demonstrated the potential effectiveness of the BAM approach for reducing indoor tanning. Follow-up analyses revealed that the intervention effects were partially mediated by changes in attitudes toward appearance-related alternatives (i.e., sunless tanning, fashion, exercise, etc.).

The IM and BAM appear to be promising theoretical models upon which to base the development of prevention interventions targeting indoor tanning behavior in young people. However, in order to be successful, it is crucial that they be implemented properly. Developing models based on the IM and BAM requires an understanding of the specific beliefs and perceptions underlying behavioral attitudes, an understanding of the behavioral alternatives perceived as practical by the target audience, and an empirical understanding of the relationships amongst all of these variables. The behavioral alternative approach, in particular, requires empirical models of the underlying beliefs motivating each alternative in order to develop effective BAM-based interventions. For example, it is not adequate to simply offer a sunless tanning session or product without understanding the underlying motivations for decisions to use these alternatives and attempting to influence these motivations. BAM-based interventions try to reduce indoor tanning motivations by targeting key beliefs related to indoor tanning, while simultaneously attempting to increase motivations toward healthy alternatives through impacting the key beliefs related to the alternatives. Developing models based on the IM and BAM require a commitment of time, effort and empirical testing to achieve efficacious results.

Results from the studies using TMT provide an explanation for reactance to health-based educational interventions [36, 37]. If cancer-related information raises anxiety in participants, indoor tanning motivation may actually increase, given the widespread presence of positive tanning images and messages in society. Interventions which focus on decreasing the positive images and appearance beliefs associated with tanning should be more successful. Changing views on the attractiveness and acceptability of pale skin may also prove useful.

On the surface, constructs from the HBM have fared inconsistently in the literature. Increased feelings of threats of skin harm or cancer have frequently been associated with increases in indoor tanning intentions and behavior. This counterintuitive finding is somewhat surprising given the model's long history of success with other health behaviors. We think it is unlikely that believing indoor tanning is a risk for cancer is causing individuals to decide to indoor tan. A more reasonable explanation is that individuals considering or currently indoor tanning are simply more aware of the indoor tanning-cancer risk due to the increased salience of the information for them. However, beliefs in cancer susceptibility from indoor tanning do not appear to be strong deterrents to indoor tanning behavior in young people. The few studies which have explored perceptions related to indoor tanning's relationship to skin aging and appearance-related damage have more consistently found negative associations with intentions and behavior. We hypothesize that examining the fit of the HBM to indoor tanning decisions would be successful if the threat, benefits, barriers, cues to action and self-efficacy variables were constructed with a focus on appearance-related issues.

Even though the SCT has never been tested fully with indoor tanning, modeling, a construct from this theory, is a consistent and potentially important variable related to indoor tanning. Parents' and peers' behaviors and beliefs have consistently demonstrated significant relationships with indoor tanning behavior. These results suggest courses of action for successful interventions directed at reducing indoor tanning behavior. Parents and peers represent a natural means of delivering intervention material to targeted individuals. Peer-led and parent-based interventions have already been proven to be effective with a number of health issues including other skin cancer risk behaviors such as reducing sunbathing and increasing sunscreen use [56, 57]. It is also possible that interventions directed at reducing indoor tanning behavior in pivotal individuals such as mothers or important social figures such as cheerleaders may have the effect of reducing tanning in the wider population. Stapleton and colleagues have data that indicate the social crowd one belongs to is an important indicator of indoor tanning decisions [58]. Intervening with leaders of the social crowds with high tanning rates may be crucial in reducing tanning throughout the group.

Interventions directed at the mothers of teens, particularly teen girls, may also prove fruitful. Baker and colleagues report that adolescents who initiated their indoor tanning with their mothers are more likely to become frequent, habitual tanners as young adults [51]. An intervention targeting mothers of young teens may reduce the mother's tanning and also have important beneficial effects on the teen too.

Bagdasarov and colleagues suggest an association between indoor tanning behavior and other risky/addictive behaviors such as drinking and smoking [34]. These findings support other research that has reported similar associations with drinking, smoking, illicit drug use, pathological weight control and other health risk behaviors [59, 60]. These results suggest a possible role for addiction motives in indoor tanning decisions and maintenance. For example, evidence for tanning dependence is presented in another chapter in this volume (Shah et al., Chap. 7) and will not be reviewed here. Such tanning motivations should be explored in relation to the established theoretical models previously studied. Understanding the relationship of such motives to these models may have important implications for the design of interventions directed at frequent, habitual tanners, an important subgroup in terms of cancer risk.

4.4 Future Directions

It is clear from this review that additional studies examining theoretical models of indoor tanning behavior are needed. In particular, studies that investigate the health belief model and social cognitive theory to a greater extent, as well as research exploring the transtheoretical model, precaution adoption process model, ecological models and other health behavior theories would be worthwhile [61]. Future work should do more comparison of theories and seek to integrate similar and complimentary constructs to develop more comprehensive models of intentional tanning decisions and behaviors. Such models will be important guides for efficacious prevention interventions. In particular, exploring social, parenting, marketing and other important behavioral influences and better integrating them into established models is suggested.

As seen in the chapter in this volume on tanning addiction, there is growing recognition that addictive processes may be important influences on indoor tanning behavior for some individuals. Such processes may lead to frequent persistent indoor tanning patterns with dire health consequences. A theoretical conceptual model of addictive tanning needs to be developed and validated in order for the utility of addiction models (e.g., genetic theories, exposure theories (both biological and conditioning), adaptation theories, learning theories,) for indoor tanning behavior to be explored.

Most but not all of the aforementioned studies are cross-sectional in design, limiting the ability to make cause and effect statements. However, the studies that have utilized a more longitudinal design together with the results of meditational analyses support the hypothesized structure of the models examined (i.e., beliefs influence attitudes and subjective norms that then influence intentions that lead to actual behavior). Future studies should use longitudinal designs to enable a better understanding of the relationships among constructs and how these constructs develop and change over time.

While stress relief/relaxation aspects of indoor tanning have been explored and found to be important predictors of indoor tanning attitudes, these parts of the models have not been fully utilized in prevention interventions to date. For example, it would be interesting to explore alternative means of stress relief using the BAM. Interventions promoting yoga, meditation, deep muscle relaxation or other healthy ways to relax may prove successful in the subgroups of indoor tanners who are strongly motivated by stress relief outcomes. Understanding the specific features of indoor tanning that provide relief of stress (e.g., warmth, solitude) would be crucial to understanding which alternatives are likely to be considered most strongly by indoor tanners.

Most studies on indoor tanning focus on young adult populations. There is a need to better understand both older and younger tanners. Theoretical models that account for developmental changes in tanning adoption and maintenance would further our understanding of the behavior and help identify crucial target populations, beliefs and behaviors as well as help us to time our interventions in a way that would give them the most impact. The fact that almost all these studies have been conducted on Caucasian samples from one culture is also an important limitation. While there is little evidence that some groups such as African-Americans indoor tan to any significant degree, there is evidence that Hispanic and Asian populations do. Future studies need to look at these other ethnic and cultural groups to determine how current models fit, and what additions or changes need to be made to better understand their tanning decisions.

Acknowledgements The authors would like to thank the American Cancer Society and the National Cancer Institute for their support of the research of Drs. Hillhouse and Turrisi and Kelly Guttman and Katie Baker for their comments on earlier versions of the chapter.

References

1. Walter SD, Marrett LD, From L et al (1990) The association of cutaneous malignant melanoma with the use of sunbeds and sunlamps. Am J Epidemiol 131(2):232–243
2. El Ghissassi F, Baan R, Straif K et al (2009) A review of human carcinogens – part D: radiation. Lancet Oncol 10(8):751–752
3. Glanz K, Maddock J (2000) On judging models and theories: research and practice, psychology and public health. J Health Psychol 5(2):151–154
4. Amir Z, Wright A, Kernohan EE et al (2000) Attitudes, beliefs and behaviour regarding the use of sunbeds amongst healthcare workers in Bradford. Eur J Cancer Care (Engl) 9(2):76–79
5. Beasley TM, Kittel BS (1997) Factors that influence health risk behaviors among tanning salon patrons. Eval Health Prof 20(4):371–388
6. Boldeman C, Jansson B, Nilsson B et al (1997) Sunbed use in Swedish urban adolescents related to behavioral characteristics. Prev Med 26(1):114–119
7. Brandberg Y, Ullen H, Sjoberg L et al (1998) Sunbathing and sunbed use related to self-image in a randomized sample of Swedish adolescents. Eur J Cancer Prev 7(4):321–329
8. Cafri G, Thompson JK, Jacobsen PB (2006) Appearance reasons for tanning mediate the relationship between media influence and UV exposure and sun protection. Arch Dermatol 142(8):1067–1069
9. Cafri G, Thompson JK, Roehrig M et al (2006) An investigation of appearance motives for tanning: the development and evaluation of the Physical Appearance Reasons for Tanning Scale (PARTS) and its relation to sunbathing and indoor tanning intentions. Body Image 3(3):199–209
10. Rhainds M, De Guire L, Claveau J (1999) A population-based survey on the use of artificial tanning devices in the Province of Quebec, Canada. J Am Acad Dermatol 40(4):572–576
11. Sjöberg L, Holm LE, Ullen H et al (2004) Tanning and risk perception in adolescents. Health Risk Soc 6(1):81–94
12. Young JC, Walker R (1998) Understanding students' indoor tanning practices and beliefs to reduce skin cancer risks. Am J Health Stud 14:120–126
13. Diffey BL (1986) Use of UV-A sunbeds for cosmetic tanning. Br J Dermatol 115(1):67–76
14. Dougherty RJ, Hawkins MJ (1988) A profile of users of commercial tanning salons. Health Values 12:21–29
15. Feldman SR, Liguori A, Kucenic M et al (2004) Ultraviolet exposure is a reinforcing stimulus in frequent indoor tanners. J Am Acad Dermatol 51(1):45–51
16. Hillhouse JJ, Stair AW, Adler CM (1996) Predictors of sunbathing and sunscreen use in college undergraduates. J Behav Med 19(6):543–561
17. Mawn VB, Fleischer AB Jr (1993) A survey of attitudes, beliefs, and behavior regarding tanning bed use, sunbathing, and sunscreen use. J Am Acad Dermatol 29(6):959–962
18. Zeller S, Lazovich D, Forster J et al (2006) Do adolescent indoor tanners exhibit dependency? J Am Acad Dermatol 54(4):589–596
19. Hoerster KD, Mayer JA, Woodruff SI et al (2007) The influence of parents and peers on adolescent indoor tanning behavior: findings from a multi-city sample. J Am Acad Dermatol 57(6):990–997
20. Flay B, Biglan A, Boruch R et al (2004) Standards of evidence: criteria for efficacy, effectiveness and dissemination. Society of Prevention Research, Falls Church
21. Ajzen I (1985) A theory of planned behavior. In: Kuhl J, Beckman J (eds) Action-control: from cognition to behavior. Springer, Heidelberg, pp 11–39
22. Fishbein M, Ajzen I (1975) Belief, attitude, intention, and behavior: an introduction to theory and research. Addison-Wesley, Reading
23. Hillhouse JJ, Adler CM, Drinnon J et al (1997) Application of Azjen's theory of planned behavior to predict sunbathing, tanning salon use, and sunscreen use intentions and behaviors. J Behav Med 20(4):365–378

24. Hillhouse JJ, Turrisi R, Kastner M (2000) Modeling tanning salon behavioral tendencies using appearance motivation, self-monitoring and the theory of planned behavior. Health Educ Res 15(4):405–414
25. Prislin R, Kovrlija N (1992) Predicting behavior of high and low self-monitors: an application of the theory of planned behavior. Psychol Rep 70(3 Pt 2):1131–1138
26. Jaccard J (1981) Attitudes and behavior: implications for attitudes toward behavioral alternatives. J Exp Social Psychol 17:286–307
27. Danoff-Burg S, Mosher CE (2006) Predictors of tanning salon use: behavioral alternatives for enhancing appearance, relaxing and socializing. J Health Psychol 11(3):511–518
28. Hillhouse J, Turrisi R, Holwiski F et al (1999) An examination of psychological variables relevant to artificial tanning tendencies. J Health Psychol 4:4507–4516
29. Hillhouse JJ, Turrisi R (2002) Examination of the efficacy of an appearance-focused intervention to reduce UV exposure. J Behav Med 25(4):395–409
30. Hillhouse J, Turrisi R, Stapleton J et al (2008) A randomized controlled trial of an appearance-focused intervention to prevent skin cancer. Cancer 113(11):3257–3266
31. Gibbons FX, Gerrard M (1995) Predicting young adults' health risk behavior. J Pers Soc Psychol 69(3):505–517
32. Gibbons FX, Gerrard M, Lane DJ et al (2005) Using UV photography to reduce use of tanning booths: a test of cognitive mediation. Health Psychol 24(4):358–363
33. Jessor R, Jessor S (1977) Problem behavior and psychosocial development. Academic, New York
34. Bagdasarov Z, Banerjee S, Greene K et al (2008) Indoor tanning and problem behavior. J Am Coll Health 56(5):555–561
35. Goldenberg JL, Arndt J (2008) The implications of death for health: a terror management health model for behavioral health promotion. Psychol Rev 115(4):1032–1053
36. Cox CR, Cooper DP, Vess M et al (2009) Bronze is beautiful but pale can be pretty: the effects of appearance standards and mortality salience on sun-tanning outcomes. Health Psychol 28(6):746–752
37. Routledge C, Arndt J, Goldenberg JL (2004) A time to tan: proximal and distal effects of mortality salience on sun exposure intentions. Pers Soc Psychol Bull 30(10):1347–1358
38. Blumer H (1969) Symbolic interactionism: perspective and method. University of California Press, Berkeley
39. Vannini P, McCright AM (2004 Summer) To die for: the semiotic seductive power of the tanned body. Symbolic Interaction 27(3):309–332
40. Fredrickson B, Roberts T (1997) Objectification theory: toward understanding women's lived experiences and mental health risks. Psychol Women Q 21:173–206
41. Stapleton J, Turrisi R, Todaro A et al (2009) Objectification theory and our understanding of indoor tanning. Arch Dermatol 145(9):1059–1060
42. McKinley N, Hyde J (1996) The objectified body consciousness scale: development and validation. Psychol Women Q 20:181–215
43. Rosenstock I (1990) The health belief model: explaining health behavior through expectancies. In: Glanz K, Lewis F, Rimer B (eds) Health behavior and health education: theory, research, and practice. Jossey-Bass, San Francisco
44. Bandura A (1989) Social cognitive theory. In: Vasta R (ed) Annals of child development, Vol 6. Six theories of child development. JAI Press, Greenwich, pp 1–60
45. Greene K, Brinn LS (2003) Messages influencing college women's tanning bed use: statistical versus narrative evidence format and a self-assessment to increase perceived susceptibility. J Health Commun 8(5):443–461
46. Coups EJ, Manne SL, Heckman CJ (2008) Multiple skin cancer risk behaviors in the U.S. population. Am J Prev Med 34(2):87–93
47. Heckman CJ, Coups EJ, Manne SL (2008) Prevalence and correlates of indoor tanning among US adults. J Am Acad Dermatol 58(5):769–780
48. Cokkinides V, Weinstock M, Lazovich D et al (2009) Indoor tanning use among adolescents in the US, 1998 to 2004. Cancer 115(1):190–198

49. Cokkinides VE, Weinstock MA, O'Connell MC et al (2002) Use of indoor tanning sunlamps by US youth, ages 11–18 years, and by their parent or guardian caregivers: prevalence and correlates. Pediatrics 109(6):1124–1130
50. Stryker JE, Lazovich D, Forster JL et al (2004) Maternal/female caregiver influences on adolescent indoor tanning. J Adolesc Health 35(6):528.e1–528.e9
51. Baker K, Hillhouse J, Liu X (2010) Does initial indoor tanning with mother impact current tanning patterns? Arch Dermatol 146(12):1427–1428
52. Lazovich D, Forster J, Sorensen G et al (2004) Characteristics associated with use or intention to use indoor tanning among adolescents. Arch Pediatr Adolesc Med 158(9):918–924
53. Cafri G, Thompson JK, Jacobsen PB et al (2009) Investigating the role of appearance-based factors in predicting sunbathing and tanning salon use. J Behav Med 32:532–544
54. Hillhouse J (2011) Modeling indoor tanning behavior. Manuscript in preparation
55. Geller AC, Colditz G, Oliveria S et al (2002) Use of sunscreen, sunburning rates, and tanning bed use among more than 10 000 US children and adolescents. Pediatrics 109(6):1009–1014
56. Turrisi R, Hillhouse J, Heavin S et al (2004) Examination of the short-term efficacy of a parent-based intervention to prevent skin cancer. J Behav Med 27(4):393–412
57. Turrisi R, Jaccard J, Taki R et al (2001) Examination of the short-term efficacy of a parent intervention to reduce college student drinking tendencies. Psychol Addict Behav 15(4): 366–372
58. Stapleton J, Turrisi R, Hillhouse J (2008) Peer crowd identification and indoor artificial UV tanning behavioral tendencies. J Health Psychol 13(7):940–945
59. Heckman CJ, Egleston BL, Wilson DB et al (2008) A preliminary investigation of the predictors of tanning dependence. Am J Health Behav 32(5):451–464
60. O'Riordan DL, Field AE, Geller AC et al (2006) Frequent tanning bed use, weight concerns, and other health risk behaviors in adolescent females (United States). Cancer Causes Control 17(5):679–686
61. Glanz K, Bishop DB (2010) The role of behavioral science theory in development and implementation of public health interventions. Annu Rev Public Health 31:399–418

Chapter 5
How Ultraviolet Radiation Tans Skin

Stuart R. Lessin, Clifford S. Perlis, and Matthew B. Zook

Abstract The choice to darken one's skin by tanning has gained increased popularity and social acceptance in recent years. The opportunities to tan intentionally have paralleled the dramatic growth of the indoor tanning industry, which asserts that tanning provides health benefits. This increase in intentional tanning has now become a public health issue. To understand why, it is important to understand how ultraviolet radiation (UVR) tans the skin. Tanning is the skin's response to UVR exposures. Interaction between UVR and our skin's pigment system is central to the tanning response. UVR exposure produces DNA damage that directly and indirectly stimulates tanning. The ability of individuals to tan is dependent upon the amount of naturally expressed pigment in their skin. Most germane to the issue of tanning, however, is that UVR-induced DNA damage produces genetic mutations that can lead to skin cancer. This chapter will delineate important biologic aspects of UVR and pigmentation and provide a better understanding of the tanning process and its role as a secondary defense against UVR-induced DNA damage.

Keywords Melanin • Melanocyte • Ultraviolet radiation – A (UVA) • Ultraviolet radiation – B (UVB) • Keratinocyte • Pigment • DNA damage • DNA repair • Skin cancer • p53 • Skin type • Evolution • Tanning • Carcinogenesis • Epidermis

Abbreviations

DNA	Deoxyribonucleic acid
nm	Nanometers
UVA	Ultraviolet radiation – A
UVB	Ultraviolet radiation – B
UVR	Ultraviolet radiation

S.R. Lessin (✉) • C.S. Perlis • M.B. Zook
Fox Chase Cancer Center, 333 Cottman Avenue, Philadelphia, PA 19111
e-mail: Stuart.Lessin@fccc.edu; Clifford.Perlis@fccc.edu; Matthew.Zook@fccc.edu

C.J. Heckman and S.L. Manne (eds.), *Shedding Light on Indoor Tanning*,
DOI 10.1007/978-94-007-2048-0_5, © Springer Science+Business Media B.V. 2012

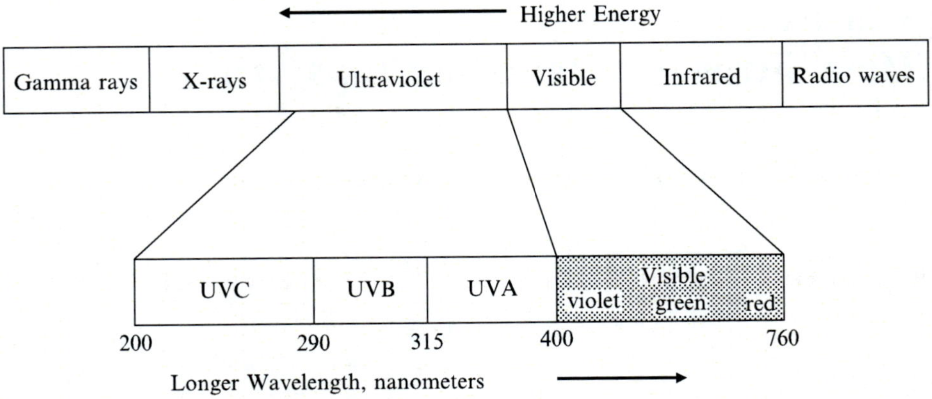

Fig. 5.1 Ultraviolet radiation wavelengths within the electromagnetic spectrum

5.1 Ultraviolet Radiation

Ultraviolet radiation (UVR) is part of the electromagnetic spectrum of energy emitted from the sun. Its wavelength extends from 10 to 400 nanometers (nm), just below the visible light spectrum (400–700 nm). Because the earth's ozone layer blocks short wavelength UVR, greater than 90% of the UVR that reaches the earth's surface is UVA (315–400 nm), and a smaller portion is UVB (290–315 nm) (Fig. 5.1). Indoor tanning devices deliver variable amounts of UVA and UVB to the skin (see Sect. 5.6).

The smaller quantities of higher energy UVB are responsible for acute sunburns of the skin, which typically develop throughout the 24 h after exposure. UVB exposures stimulate an increase in pigment production in the skin that is responsible for tanning. UVA penetrates deeper into the skin and does not cause acute sunburns but causes an increase in pigment production and tanning up to 72 h after exposure. Both UVA and UVB cause a reaction of the cellular DNA repair mechanisms that are closely linked and highly coordinated with pigment production.

5.2 DNA Damage and Skin Cancer Development

DNA damage and its repair pathways are part of the initial steps in stimulating pigmentation and tanning. UVR-related carcinogenesis (cancer development) in the skin is initiated when skin cell DNA absorbs UVR energy. The ringed-structured bases of DNA absorb strongly in the UVB wavelength, but UVA can also trigger DNA damage [1, 2]. UVR energy absorption by DNA generates abnormal molecular cross-links between DNA base pairs (nucleotides) that create genetic mutations if left unrepaired [2, 3].

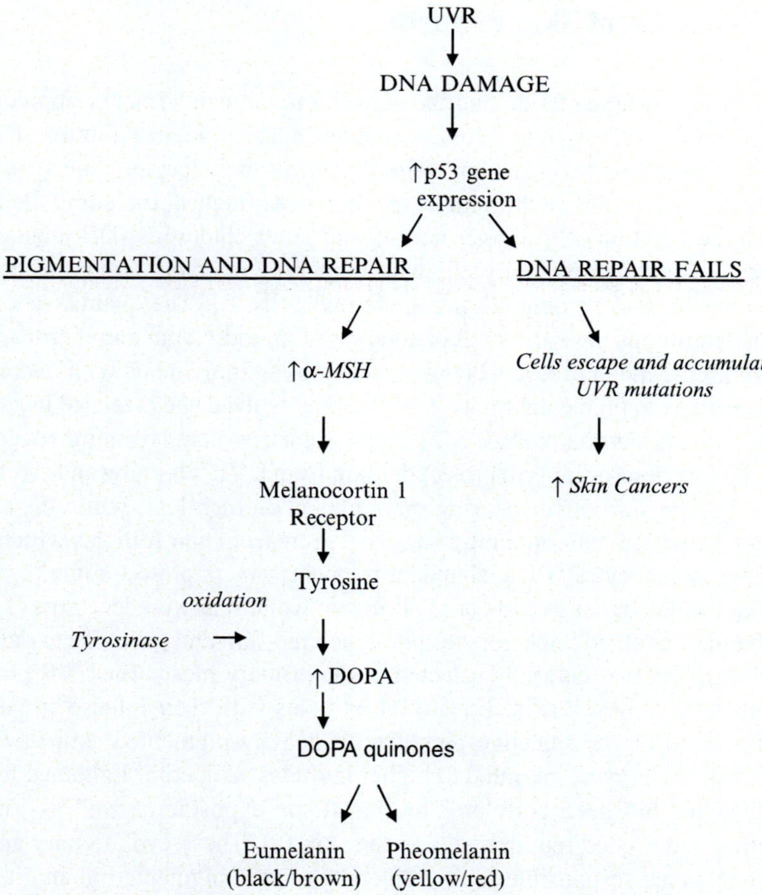

Fig. 5.2 Molecular steps in tanning and UVR-induced carcinogenesis. Tanning is initiated as a response to UVR-induced DNA damage in the skin. An early response gene is p53, which plays a central role in DNA repair. Cells that escape DNA repair pathways accumulate UVR-related mutations leading to skin cancer formation. UVR-induced DNA damage is linked to the pigmentation response through p53-dependent induction of *a*MSH which acts on melanocortin 1 receptors (MC1R) on melanocytes (agouti signaling protein (ASIP) also acts on MC1R, see Sect. 5.4). MC1R engagement (and p53) induces tyrosinase activity and the oxidation of tyrosine, the building block of melanin. Subsequent synthesis of dopaquinone is further modified into either eumelanin or pheomelanain

The p53 tumor suppressor gene is central to the DNA repair response, and its activation is linked to pigmentation and tanning [3, 4] (Fig. 5.2). p53-mediated cell cycle arrest provides time for elimination of UVR-induced DNA mutations in the skin cells by the mechanism of nucleotide excision repair [3, 4]. In cells overwhelmed by DNA damage beyond repair, p53 mediates apoptosis (programmed cell death) and the elimination of these cells from the skin. UVR-induced skin cancers arise when cells escape from the DNA repair pathways and accumulate a critical threshold of mutations leading to malignant transformation. For more on skin cancer, please refer to Chap. 6 of this book by Zook and colleagues.

5.3 Evolution of Skin Pigment

The diversity of pigmentation and the capacity to tan can be better appreciated by an understanding of how skin color has changed during human evolution. The skin's pigment is called melanin, which is produced by melanocytes, one type of skin cells. These cells populate the skin in specific areas (niches) including the hair follicle and the junction of the inner dermis and outer epidermis (DE junction). The melanocyte is a dendritic cell (cells with branching intercellular connections) and provides melanin to its neighboring epidermal cells [5]. The quantity and type of melanin determines the color of skin and hair of an individual and their capacity to tan, with higher melanin levels being present among individuals with darker skin.

The diversity in pigmentation we see today has evolved and is related to adaptation to changes in the environment [3, 6, 7]. Homo sapiens evolved from fur covered apes. One of the functions of fur is to protect the skin from UVR's burning and carcinogenic effects [7]. The skin of fur-bearing mammals is pigmentless, while the colors of fur result from pigment-containing hairs that grow from hair follicles with melanin-containing melanocytes. Thus, abundant pigmented hairs protected the pigmentless skin from the deleterious effects of UVR in our evolutionary predecessors [7].

As humans evolved, melanocytes populated the skin and gave rise to darkly pigmented skin that was naturally selected as the primary means for UVR protection, supplanting fur-covered skin. Equatorial latitudes with their intense sun exposure are believed to be the selective pressure for heavily-pigmented skin first seen in humans [3]. As humans migrated to higher latitudes, skin color lightened following a gradient that has been correlated to annual sun exposure as well as Vitamin D production and related requirements in the skin [8]. These evolutionary and selective pressures are responsible for the diversity of skin pigmentation and its current role in skin biology [3, 8].

5.4 Physiology of Pigmentation

Within melanocytes, melanin is produced in membrane-bound organelles (cell subunits with specialized functions) termed melanosomes. There are two types of melanin produced in melanosomes. These include eumelanin that confers brown to black pigment and pheomelanin that confers yellow to red pigment. The amino acid tyrosine is the common building block for both melanins. Tyrosinase is the first enzyme in a complex synthetic pathway that oxidizes tyrosine to dopaquinone [9]. Further enzymatic modifications of dopaquinone represent a branch point in melanogenesis from which eumelanin or pheomelanin is synthesized (Fig. 5.2). Skin's melanin content consists of a mixture of both pigments that contribute to the capacity to tan and to the risk of skin cancer [9, 10] (see Sect. 5.5).

Melanin-containing melanosomes are intercellularly transferred from melanocytes into epidermal cells (keratinocytes) through cellular extensions (dendrites).

The exact mechanism is unknown, but there appears to be a precise ratio (epidermal melanin unit) of melanocytes to keratinocytes throughout the epidermis responsible for the maintenance of the skin's pigmentation [5]. Melanosomes aggregate over keratinocyte nuclei (cell control centers), in a process known as nuclear capping, where they shield basal keratinocyte nuclei from UVR [5]. This intra-cellular sunscreen is vital in maintaining the renewable layer of the epidermis.

The constitutive quantities of melanin that determine an individual's skin color are controlled through specific surface receptors on melanocytes. Melanocortin-1 receptor (MC1R) binds melanocortin or *a*-melanocyte stimulating hormone (*a*MSH) that stimulates the production of eumelanin [10, 11]. Agouti signaling protein (ASIP) also binds to MC1R and stimulates pheomelanin formation [10, 11]. Genetic variations in the MC1R's capacity to respond to MSH or ASIP stimulation influences cellular levels of eumelanin and pheomelanin and the resultant skin color of each individual [10, 11].

5.5 Physiology of Tanning

Tanning is a response to UVR-induced DNA damage. UVR stimulates a wide array of genes and signaling pathways involved with pigmentation and skin cancer susceptibility. Tanning is a highly-coordinated response directed at repair and reduction of UVR-induced DNA mutations.

UVR triggers the expression of DNA nucleotide excision repair genes with the capacity to specifically repair UVR signature mutations. Additionally, UVR exposure leads to the production of a wide array of cytokines within the skin. These cytokines are both pro-inflammatory and melanogenic (melanin-producing). p53 gene expression coordinates the DNA repair pathways and drives melanogenesis through the induction of the precursor compounds of *a*MSH and pro-opiomelanocortin (POMC) in keratinocytes [4] (Fig. 5.2). Photo-damaged pyrimidine dimers, excised during DNA repair, induce tyrosinase expression [12] as does p53 [13]. Both drive increased melanin production and stimulate quantitative and qualitative changes in melanin.

Increased pigment synthesis within melanocytes results in increased melanosomes and transfer of newly-formed melanosomes to neighboring keratinocytes. The UVR-induced increase and distribution of melanin throughout the basal layer of keratinocytes within the epidermis is what is observed as tanning; however, the principle effect of this process is to reduce DNA damage by forming a protective shield of UVR-absorbing melanin over keratinocyte nuclei (Fig. 5.2).

An individual's capacity to tan is related to his/her pigment-producing capacity and is clinically categorized by Fitzpatrick skin type that classifies skin by skin color, susceptibility to burning, and ability to tan [14] (Table 5.1). Fair complexions are considered skin type I and II typified by light-skinned, blond or red-haired individuals who have predominately pheomelanin and burn readily after UVR

Table 5.1 Fitzpatrick skin types (Adapted from [12])

Type	Appearance	Sun sensitivity	Reaction to sun
I	Red-haired; freckled	Sensitive	Always burns easily, never tans
II	Fair-skinned; blue eyes	Sensitive	Always burns easily, tans minimally
III	Dark-skinned white	Normal	Burns moderately, tans gradually
IV	Mediterranean	Normal	Burns minimally, tans always
V	Middle-Eastern, Latin-American, light-skinned African-American	Little to none	Rarely burns, tans profusely
VI	Dark-skinned African-American	Little to none	Never burns, deeply darkens

exposures and rarely, if ever tan. With increasing amounts of eumelanin, complexions are darker as seen in dark-skinned Caucasians and Mediterranean whites (skin types III–IV). The most heavily pigmented skin is seen in Hispanics and Africans (type V–VI).

The higher the skin type, the longer an individual can tolerate UVR exposure without burning, but all skin types burn from UVB exposures. Repeated exposures result in tanning that does provide a degree of sun protection for all skin types but also stimulate melanogenesis [15]. Melanin production will return to baseline levels without repeated exposures to UVR.

5.6 Indoor Tanning

Indoor tanning is achieved with the use of UVR light bulbs that emit predominately UVA, like the sun. While current regulations permit no greater than 5% UVB, the exact percentage can vary [16]. UVA produces DNA damage and stimulates pigmentation [1, 2], but its longer wavelength and lower energy are less likely to cause the acute burning seen with UVB overexposures. Most commercial tanning devices utilize multiple long fluorescent bulbs that predominately emit UVA and smaller varying amounts of UVB. The bulbs are aligned in either a vertically-oriented enclosure (tanning box or booth) or a horizontal enclosure (tanning bed). "High-pressure" tanning devices utilize smaller quartz bulbs manufactured with high internal pressures. These bulbs require filtering systems to eliminate its UVB sources. High-pressure tanning bulbs emit as much as 99–100% UVA radiation in order to increase the duration of tans while decreasing the likelihood of some negative effects of UVB such as skin burning and dryness [16]. The longer wavelength UVA penetrates deeper into the skin than UVB and stimulates melanogenesis and pigmentation; however, compared to UVB, UVA exposures induce a delayed tanning effect. Indoor tanning devices therefore require multiple and repeated UVA exposures to tan the skin. Indoor UVA exposures have the same biologic effect and are as deleterious to the skin as UVA exposure from sunlight [3, 16].

5.7 Summary

UVA is predominantly responsible for tanning of the skin, and UVB is responsible for burning. Tanning is a human defense mechanism directed against DNA damage induced by UVR exposures. Melanin-containing melanocytes are closely associated with keratinocytes within the epidermis, and baseline pigmentation is tightly controlled by complex cellular and intercellular pathways. Evolutionary and adaptive changes have created the diversity of human pigmentation, skin types, and tanning capacities. Tanning results when UVR-induced melanogenesis increases pigment levels in the skin above baseline, and repeated exposures maintain these elevated levels. Tanning is a secondary response to UVR-induced DNA damage and molecularly linked to DNA repair pathways. Tanning beds produce UVR similar to the sun, and likewise cause DNA damage and contribute to the development of skin cancers among humans.

References

1. Ravant JL, Douki T, Cadet J (2001) Direct and indirect effects of UV radiation on DNA and its components. J Photochem Photobiol B 63:88–102
2. Mouret S, Baudouin C, Charveron M et al (2006) Cyclobutane pyrimidine dimers are predominant DNA lesions in whole human skin exposed to UVA radiation. Proc Natl Acad Sci USA 103(37):13765–13770
3. Schulman JM, Fisher DE (2009) Indoor ultraviolet tanning and skin cancer: health risks and opportunities. Curr Opin Oncol 21(2):144–149
4. Cui R, Widlund HR, Feige E et al (2007) Central role of p53 in the suntan response and pathologic hyperpigmentation. Cell 128(5):853–864
5. Singh SK, Kurfurst R, Nizard C et al (2010) Melanin transfer in human skin cells is mediated by filopodia – a model for homotypic and heterotypic lysosome-related organelle transfer. FASEB J 24(10):3756–3769
6. Juzeniene A, Setlow R, Porojnicu A et al (2009) Development of different human skin colors: a review highlighting photobiological and photobiophysical aspects. J Photochem Photobiol B 96(2):93–100
7. Jablonski NG, Chaplin G (2000) The evolution of human skin coloration. J Hum Evol 39(1):57–106
8. Yuen AW, Jablonski NG (2010) Vitamin D: in the evolution of human skin colour. Med Hypotheses 74:39–44
9. Ito S, Wakamatsu K (2008) Chemistry of mixed melanogenesis – pivotal roles of dopaquinone. Photochem Photobiol 84(3):582–592
10. Flanagan N, Healy E, Ray A et al (2000) Pleiotropic effects of the melanocortin 1 receptor (MC1R) gene on human pigmentation. Hum Mol Genet 9(17):2531–2537

11. Le Pape E, Passeron T, Giubellino A et al (2009) Microarray analysis sheds light on the dedifferentiating role of agouti signal protein in murine melanocytes via the Mc1r. Proc Natl Acad Sci USA 106(6):1802–1807
12. Eller MS, Ostrom K, Gilchrest BA (1996) DNA damage enhances melanogenesis. Proc Natl Acad Sci USA 93(3):1087–1092
13. Khlgatian MK, Hadshiew IM, Asawanonda P et al (2002) Tyrosinase gene expression is regulated by p53. J Invest Dermatol 118(1):126–132
14. Fitzpatrick TB (1988) The validity and practicality of sun-reactive skin types I through VI. Arch Dermatol 124(6):869–871
15. Sheehan JM, Cragg N, Chadwick CA et al (2002) Repeated ultraviolet exposure affords the same protection against DNA photodamage and erythema in human skin types II and IV but is associated with faster DNA repair in skin type IV. J Invest Dermatol 118(5):825–829
16. Woo DK, Eide MJ (2010) Tanning beds, skin cancer, and vitamin D: an examination of the scientific evidence and public health implications. Dermatol Ther 23(1):61–71

Chapter 6
Skin Cancer and Other Health Effects of Indoor Tanning

Matthew Zook, Stuart Lessin, and Clifford Perlis

Abstract Research clearly demonstrates the adverse health effects of ultraviolet radiation (UV) from the sun on human skin. Among these adverse effects, photoaging and the development of skin cancer are perhaps the most clinically important. There is a large effort directed towards combating these problems and the resultant morbidity and mortality associated with them, particularly the increased risk for skin cancers. Indoor tanning provides ultraviolet radiation exposure, which is similar to that of the sun. Indoor tanning has increased dramatically over the last two decades. Unfortunately, we are finding that indoor tanning leads to similar, if not worse, adverse health effects than the sun. In this chapter, the effects of ultraviolet radiation on skin cancer development, photoaging, the immune system and other medical conditions are discussed in the context of indoor tanning.

Keywords Skin cancer • Melanoma • Squamous cell carcinoma • Basal cell carcinoma • Artificial tanning • Tanning bed • Photoaging • Phototherapy • Ultraviolet light • Immunity • UVA • UVB • Photodynamic therapy • International Agency for Cancer Research • World Health Organization

Abbreviations

UV	Ultraviolet radiation
FDA	Food and Drug Administration
BCC	Basal cell carcinoma
SCC	Squamous cell carcinoma
PUVA	Psoralen and ultraviolet A light

M. Zook (✉) • S. Lessin • C. Perlis
Fox Chase Cancer Center, 333 Cottman Avenue, Philadelphia, PA 19111, USA
e-mail: Matthew.Zook@fccc.edu; Stuart.Lessin@fccc.edu; Clifford.Perlis@fccc.edu

C.J. Heckman and S.L. Manne (eds.), *Shedding Light on Indoor Tanning*,
DOI 10.1007/978-94-007-2048-0_6, © Springer Science+Business Media B.V. 2012

6.1 Background

The fact that ultraviolet (UV) light can affect the skin has been known since antiquity. It is believed that the ancient Egyptians used sun exposure in an attempt to treat conditions such as vitiligo (loss of skin color) [1]. Conversely, the inability of those with porphyria (a defect in heme biosynthesis) to tolerate the sun, along with the excessive hair growth that is seen in this disorder, lead to the modern day fairy-tale concept of the werewolf [2]. Fortunately, modern society has moved beyond this primitive understanding of the effects of UV light.

Over the past 200 years, the general population's perceptions of the effects of sun exposure have changed. Prior to the industrial revolution, those with power and affluence in the Western world generally had fair, untanned skin because they were Caucasian and could avoid outdoor physical labor. This propagated a view of a tanned appearance as being associated with lower socioeconomic status. With the development of the modern factory and the subsequent migration of jobs to an indoor setting, this dichotomy was reversed. With this reversal, tanned skin signified travel, outdoor leisure activities, and a higher socioeconomic status. This view was propagated in the US and Europe in the late 1920s due to the fashion industry's acceptance of tans as chic and continued as social norms changed in the mid twentieth century [3]. The end result is a societal perception that a tan correlates with success and affluence, as well as enhancing beauty and health. This perception has spawned an industry that is centered on achieving a tanned appearance. For a more comprehensive description of the history and culture of tanning, see Chap. 2 of this book by Hunt and colleagues.

There has been a lengthy disagreement between the medical profession and the tanning industry regarding the risk of cutaneous malignancy (cancerous skin cell growth) and its correlation with exposure to artificial ultraviolet radiation. The medical profession has demonstrated that ultraviolet light causes changes in the skin such as mutations in DNA which can be mutagenic (cancer-causing) [4]. However, UV light also has been used as a treatment for skin conditions such as psoriasis, acne, and eczema, and assists in the conversion of 7-dehydrocholesterol into cholecalciferol, an early step in the synthesis of Vitamin D (see Chap. 8 by Mendese and Gilchrest). In this chapter, we will present data regarding UV light and artificial tanning and their role in skin cancers such as melanoma, basal and squamous cell carcinomas, and other health effects including photoaging and immunosuppression.

6.2 Artificial Tanning and the Risk for Melanoma

Melanoma is the third most common form of skin cancer, but the most feared due to its potential for metastasis (spreading of cancer cells). According to National Cancer Institute Surveillance Epidemiology and End Results data, 68,131 people will develop melanoma and 6,800 will die of their disease in 2010 [5]. This figure translates into one death per hour due to this cancer. In approximately 10% of the cases, there is a known genetic component, whether it is a specific mutation (for example a mutation in the tumor suppressor gene CDKN2A) or a trend among family members to be susceptible to the development of melanoma [6]. In the other 90% of cases, melanoma appears to be sporadic, being caused by a convergence of biological, behavioral and environmental risk factors [7]. There is little that can be done to change a person's biological risk factors, and changing the environment a person lives in is often not feasible. We can, however, change UV exposure behavior that places individuals at increased risk for skin cancer. Thus, it may be helpful to examine the data linking UV radiation with melanoma.

There are several risk factors for development of melanoma. Individuals with a personal or family history of melanoma have up to a ninefold increased relative risk compared to the general population [8]. Individuals with 50 or more moles or any number of atypical (irregular borders, color, or shape) moles have a 3- or 27-fold increased relative risk, respectively [9]. People with fair-colored skin who cannot respond to UV exposure by developing darkened skin color (i.e., cannot tan) are also at an increased risk of melanoma, although the level of risk is more difficult to quantify [8]. Since the majority of the population has been exposed to the sun in varying degrees and the effects of sun exposure can take decades to develop, it is difficult to provide a precise estimate of risk from sun exposure. Additionally, it would be unethical to conduct randomized controlled trials involving exposing people purposely to a carcinogen (ultraviolet radiation) in an effort to produce a cancer. Therefore, the link between UV exposure and melanoma was, until recently, anecdotal. This has made it difficult in the past to formulate a strong argument supporting the link between melanoma and UV exposure.

Several facts point towards a link between ultraviolet radiation exposure and melanoma. Those who are physically unable to tan, have fair skin with red or blond hair, or sunburn easily have a higher risk for melanoma [10]. In Caucasians, the risk for the development of melanoma is inversely related to geographic latitude, suggesting that increased ultraviolet radiation exposure is responsible for this difference [11]. This association of higher levels of UV with skin cancer development is most evident in Australia, an equatorial country populated by a large cohort of whites from Europe, which has the highest rate of melanoma in the world [12]. Finally, individuals who are unable to repair the type of DNA damage caused by ultraviolet radiation exposure are at increased risk for melanoma. In an elegant review, Gilchrest and colleagues examined these findings and others in the context of the response of melanocytes (pigmented skin cells) to ultraviolet radiation exposure [13]. The authors suggested that melanomas arise from intermittent intense exposure due to the fact that melanocytes have a low propensity to undergo apoptosis (cell death) when UV-induced damage to proteins or DNA occurs. Repeated UV exposure then leads to accumulation of DNA mutations which can lead to malignancy.

There were two early studies cited widely by skeptics of the melanoma – UV exposure link. In the first study, Osterlind and colleagues assessed a Danish population over a 3-year period [14]. While there was an association between sunbathing and melanoma among those with a history of sunburn before age 15, there was no association between the use of tanning beds and melanoma. In the second study,

Swerdlow and Weinstock analyzed the data available in 1998 and concluded that there was not enough information to prove an association between artificial ultraviolet light and melanoma [15]. Both studies have been used by the tanning industry to bolster the argument that melanoma is not due to artificial UV exposure. The tanning industry purchased advertising space in conspicuous venues such as *The New York Times* in 2008 presenting these findings. The advertisement contained a message that indoor tanning was not proven to cause melanoma and that there were health benefits such as vitamin D production which would be missed if people avoided exposure. It is not clear what impact this advertisement had on the general population because the media impact of these advertisements on tanning attitudes or behavior was not reported.

After those early reports in the 1980s and 1990s, additional studies were performed. However, there were a number of methodological challenges to overcome. The first issue is the use of retrospective recall of tanning behavior and sun exposure (e.g., sunburns) which may have occurred in childhood (which, in some cases, is more than 30 years prior to the study assessment). Second, it is difficult to compare the findings of one study with another when different inclusion criteria and endpoints are often used. For example, one study may include anyone who has ever indoor tanned, whereas another may include those who tanned prior to age 35. In spite of these methodological challenges, a great deal of evidence has accumulated to support an association between artificial UV exposure and melanoma.

One of the more comprehensive reviews was performed by the International Agency for Cancer Research, an arm of the World Health Organization, in 2007 [16]. The authors identified 23 articles which addressed the relationship between melanoma and tanning beds. Nineteen studies that included relative risks (the remainder were excluded because they did not include relative risks) were used to compile the findings using data from 7,355 individuals. Fifteen studies included European countries, two studies were from the United States, and one study each was from Canada and Australia, reporting on their respective populations. Results indicated that individuals who had ever used tanning equipment in their lifetime had a 1.15 relative risk of developing melanoma compared to the general population. This difference was statistically significant. In other words, those who had ever used a tanning bed had a roughly 15% increased chance of developing melanoma. Additionally, those that had used a tanning bed before age 35 had a relative risk of 1.75, which was also statistically significant. The risk for melanoma was tripled among those who had used two or more types of tanning beds such as walk in and table-top models. There was no correlation between melanoma risk and year of publication among articles reviewed by the IARC, which suggests that newer types of beds are not safer than older ones.

In 2009, the IARC published new evidence-based monographs updating the list of known cancer-causing agents [17]. In these reports, UV radiation was reclassified as "carcinogenic to humans", a move from the previous classification of "probably carcinogenic". This prompted the Food and Drug Administration (FDA), which is tasked with overseeing medical devices, to reexamine their classification of tanning beds. At a specially-convened task force meeting in 2010, the FDA panel recommended reclassifying tanning beds from class I to class II devices [18]. Class II devices include X-ray machines and wheelchairs, and the proposed changes would mean that tanning beds would require close inspections, calibration, and monitoring. At the time of publication of this book, it remains to be seen what position the FDA will ultimately take and what effect this may have on the use of tanning beds. However, considering reclassifying tanning beds is widely felt to be a step in the right direction as evidenced by position statements from the American Academy of Dermatology and the American Cancer Society which state that tighter regulations will help to decrease the highest risk use [19, 20]. For more information on policy, regulation, and enforcement of the tanning industry see Chap. 10 by Suarez and colleagues.

6.3 Artificial Tanning and Non-melanoma Skin Cancers

Basal cell carcinoma (BCC) and squamous cell carcinoma (SCC), malignancies of keratinocytes (skin cells), represent the majority of skin cancers in the United States and are among the most common of all cancers [21]. Squamous cell carcinoma has a greater metastatic potential than BCC, albeit lower than for melanoma. However, due to the high prevalence of SCC, it remains a significant cause of mortality, causing approximately 8,000 cases of metastatic spread to lymph nodes and resulting in approximately 3,000 deaths annually [22]. Similarly, due to the sheer incidence, the morbidity of both SCC and BSS is large, accounting for hundreds of millions of dollars of health-care expenditures yearly.

Squamous cell carcinomas are thought to correlate directly with chronic UV exposure [23], as opposed to BCCs that are widely reported to correlate with intermittent acute UV exposure [24]. This is perhaps due to a dose saturation effect noticed for basal cell carcinomas in which higher levels of UVB do not lead to an increased number of cancers [25]. Squamous cell carcinomas are known to be associated with mutations in p53 (a tumor suppressor gene), which has been found in over 90% of these cancers [26]. p53 mutations are known to be associated with UV exposure [27]. When these mutations occur, cellular clones are unable to initiate apoptotic (programmed cell death) pathways, and they proliferate abnormally, the initial steps of oncogenesis (the development of cancer cells). Similarly, high levels of COX-2 (cyclooxygenase-2), an enzyme involved in the inflammatory response, have been found in SCCs and have been thought to promote tumorigenesis (tumor development) by decreasing apoptosis. BCC is also known to carry the p53 mutations, although to a lesser extent, and BCC has been found to have mutations in PTCH1, another tumor suppressor gene [24]. When mutations are present in this pathway, cells can divide abnormally. UV induced mutations in this pathway have been found in up to 50% of sporadic BCCs [24].

Specific reviews regarding the risk for BCC or SCC induced by artificial UV radiation are not as prevalent as those for melanoma. Over the last decade, there have been nine studies that investigated the association between SCC or BCC and artificial tanning. Several studies provide relative risks [16]. When examined as a group, the studies showed that those who had ever used a tanning bed had a 2.25 relative risk for SCC and a 1.03 relative risk for BCC when compared to the general population [16]. The risk for SCC was statistically significant and indicates that individuals who have used a tanning bed are more than twice as likely to develop SCC as those who haven't tanned indoors. A thorough review by Karagas and colleagues showed that the risk for either cancer increased with tanning bed exposure [28]. The relationship was inversely proportional to the age at which the exposure took place. For example, a person who began using a tanning bed in their 20s was 20% more likely to develop a SCC and 10% more likely to develop a BCC than someone who started using a tanning bed in their 30s.

6.4 Photoaging

The term photoaging was first used in 1986 to describe the visible effects of ultraviolet radiation on the skin [29]. However, the phenomenon had been described as an effect of UV light on outdoor workers in the nineteenth century [30]. Changes to the skin associated with photoaging include dryness, hyperpigmentation (dark areas), wrinkling, telangiectasias (visible blood vessels), laxity (loss of firmness) and elastosis (damage to the non-cellular components of the skin) [31]. Whereas chronologically aged skin is atrophic, having lost its fullness, photoaged skin may be hyperplastic or thickened. Hyperplasticity is thought to be a protective response, which can shield lower layers of the epidermis (outer layer of the skin) from further UV damage.

Additionally, when skin becomes hyperplastic, collagen becomes disorganized [32], and solar elastosis forms. Solar elastosis is a grayish monomorphous substance found in the upper dermis (inner layer of the skin) of chronically sun-exposed skin and is thought to result from three processes: (1) altered degradation of existing dermal components, (2) altered synthesis of elastin, fibrillin, and glycosaminoglycans, which are protein and carbohydrate molecules that make up the structure of the skin [33], and (3) an imbalance in the regulation of collagen production [31]. Elastosis occurs through decreasing levels of signals that promote collagen formation, such as transforming growth factor, and increasing levels of components that inhibit the amount of collagen, such as matrix metalloproteinases. The end result is a continual loss of collagen, which causes the skin to lose its fullness and form wrinkles. In one of the few studies of this process Lavker and colleagues proved that these changes can be seen not only in skin exposed to the sun, but also in skin exposed to artificial ultraviolet radiation [34]. In the study, volunteers were exposed to artificial UVA and were found to have epidermal thickening, depletion of Langerhans cells (see section below on immune effects), and changes in elastin fibers as compared to volunteers who were not similarly exposed.

In addition to the changes in dermal composition, melanocyte homeostasis is altered in photaged skin, resulting in freckles, liver spots, and increased numbers of moles.

As part of normal aging, the number of melanoctyes decreases by approximately 10% per decade in adult life [35]. However, in photoaged skin, the melanocyte number may increase. The results of photoaging are typically seen as freckling. Freckling results from melanocytes becoming slightly enlarged, increasing in number, and becoming more active than usual [36]. Similarly, solar lentigos, or liver spots, develop in those habitually exposed to ultraviolet light [37]. Lentigos consist of increased numbers and activity of melanocytes as well as alterations at the junction between the dermis and epidermis. Finally, melanocytic nevi (moles), collections of melanocytes at the dermo-epidermal junction, are also seen in ultraviolet-exposed skin. Moles typically begin to occur in childhood and have been statistically associated with sun exposure [38].

6.5 UV Radiation and the Immune System

Thirty years ago, scientists discovered that UV exposure could impair the immune system [39]. When tumors induced by UV exposure were taken from one mouse and transplanted into other animals, which were not exposed to UV light, the tumors did not grow. When the recipient mice were treated with UVB or immunosuppressive agents, the tumors grew. These results showed that UV-induced tumors can be recognized and removed by the immune system in healthy animals. Conversely, the results also showed that UV exposure can decrease the ability of an animal to mount this type of immune response. Since that time, many, but not all, of the factors responsible for the immunomodulatory effects of UV radiation have been determined.

UV radiation affects the immune system through its effect on chromophores, which are portions of a molecule that absorb light. Both DNA and urocanic acid, a molecule found in the stratum corneum (the top layer of the epidermis), have been shown to be important in affecting the immune system [40]. UV radiation causes mutations in the strands of DNA. These mutations are subsequently recognized by endonucleases (enzymes), which are responsible for fixing these changes prior to cellular division [41]. As endonucleases begin the process of repairing DNA, they also cause an increase in molecular messengers, such as TNFα and IL-10, which influence the way the immune system works. Specifically, these messengers favor the formation of antibodies that are not effective in rejecting tumors [40]. The end result is that ultraviolet radiation, in addition to causing changes that lead to cancer formation, also makes the skin less likely to fight the cancers when they do develop. In a similar manner, trans-urocanic acid (a specific type of urocanic acid) is converted to cis-urocanic acid upon UV exposure, which has been shown to be immunosuppressive [42]. Removing the cis component of the stratum corneum has been found to counteract UVB-induced immune suppression [43]. Therefore, when UV radiation increases the amount of cis-urocanic acid, the ability of the immune system to recognize and respond to UV induced damage is hindered.

Changes in immune cell number and function have been noted after UV exposure, due in large part to the processes mentioned above. Epidermal Langerhans cells, phagocytic cells responsible for engulfing infectious agents and initiating an immune response, are decreased in UV-exposed skin due to migration away from the skin and through increased apoptosis [43]. Additionally, the ability of Langerhans cells to present engulfed antigen to T cells is reduced after UV exposure. Both these changes decrease the activation of the immune system, resulting in an inability of the skin, and the body overall, to protect itself from disease. UVB has also been found to cause phenotypical changes (changes in the characteristics) in macrophages causing them to preferentially secrete IL-10 [44], which as mentioned above, means the body cannot effectively fight off cancers when they develop. Similarly, neutrophil production of the immunosuppressive messenger molecule IL-4 is induced by UVB [45]. Finally, keratinocytes themselves produce TNFα, IL-10 and prostaglandin E2 in response to UV radiation in mice [43]. Taken together, these changes effectively down-regulate local and systemic immunity. In general terms, this means that the skin and body are not able to effectively guard against infections and the formation of cancers.

While we can assume that the above mentioned changes occur with tanning bed use, a specific decrease in the body's ability to repair DNA damage after tanning bed use has been reported [46]. This supports the idea that any type of UV exposure can alter the immunity of a person and lead to undesirable effects such as higher risk for cancers.

6.6 Medical Use of Artificial UV Radiation – Phototherapy

As stated at the opening of the chapter, UV has been known to provide some health benefits. As our understanding of the role of UV in treating certain skin conditions as well as the technology to deliver UV to the skin with less risk has improved, the use of phototherapy has also grown. Phototherapy, the therapeutic use of light to treat medical conditions, is delivered by UV light boxes used under careful medical supervision. The use of narrow-band UVB boxes, which emit wavelengths 311–312 nm of light, has been shown to be an effective treatment for psoriasis (red, irritated skin) [47]. Additionally, UVA boxes, which emit from 340 to 400 nm of light, have become more prevalent and have been shown to be useful for skin disorders such as scleroderma (a disease of the connective tissue that results in a variety of skin changes) [48]. This represents expansion of phototherapy as a viable option for conditions previously not considered for UV treatment.

The combined use of psoralen, a photosensitizer which makes the skin more sensitive to UV light, in combination with UVA, is known as PUVA. This treatment has been used widely for a variety of skin conditions including malignancies such as mycosis fungoides (a type of lymphoma that occurs in the skin) as well as inflammatory disorders such as psoriasis, eczema (dry, flaky skin), and graft-versus-host disease, in which the skin is attacked by immune cells. Methoxsalen, or 8-MOP, is the only type of psoralen available in the United States; bergaten and trioxsalen are available in Europe [49]. These compounds function by intercalating, or inserting themselves, between DNA base pairs and causing cyclobutane dimers (a type of incorrect pairing of two subunits of DNA) when exposed to UVA. This dimerization causes suppression of mitosis (cell division) and a slowing of the cell cycle. This treatment is therapeutic in diseases in which the skin is hyperproliferative (such as psoriasis), contains proliferating clones (such as mycosis fungoides), or is inflamed (such as eczema). Psoralens are available in oral or topical form, and when combined with UVA in a treatment regimen, are given three to five times per week. PUVA is generally thought to be an effective treatment for widespread disease. However, its use is hindered by the schedule, which includes treatments several times per week. There is also a risk for development of skin cancers. Light-complexioned patients who had received more than 260 PUVA treatments were found to have had an 11-fold higher risk of SCC than those who had less than 160 PUVA treatments [50].

Narrow band UVB (nbUVB, with a wavelength of 311 nm) has emerged as an appealing alternative to PUVA due mainly to the fact that it is thought to have less carcinogenic potential. It can also be used with lower risk in pregnancy and childhood and does not require additional post-treatment sun protection as does PUVA. The treatment regimen is roughly similar to that used for PUVA, consisting of two to five treatments per week.

More recently, photodynamic therapy (PDT) has emerged as an additional treatment option for non-melanoma skin cancers and pre-cancerous actinic keratoses. This treatment modality is similar to PUVA and UVB phototherapy in that light is used to activate a photosensitizer, in this case aminolevulinic acid (ALA). Because there is greater uptake of ALA in cancerous and pre-cancerous cells, greater amounts of reactive oxygen species (molecules with unpaired oxygen subunits that can cause damage to cells) are generated in the targeted cells when exposed to the light, which causes a cytotoxic (toxic to the cells) effect [51]. The difference between PDT and the other phototherapies lies in the type of light used; PDT uses visible light in the blue or red range, rather than UV light. Thus, there is no additional risk of skin cancer.

It is important to keep in mind the difference between medical phototherapy units and commercial tanning beds. Medical units are designed to deliver precise amounts of defined wavelengths in an effort to reach a specific endpoint (improvement of a disease). Medical units are regulated by the FDA as medical devices and are only allowed to be used under the supervision of a medical doctor. These safeguards are in place in order to ensure patient safety since medical units have the same side effect profile as commercial units, specifically exposure to carcinogenic ultraviolet radiation. The difference lies in the fact that with the use of medical units, the exposure is closely monitored and is considered necessary in order to treat a disease. Specific guidelines have been developed to aid in appropriate use of these devices [52]. With these controls in place, the devices are a useful adjunct for the treatment of several skin diseases.

6.7 Summary

It is clear that UV radiation is a powerful mediator of biologic processes in human skin. Intense or chronic overexposure from natural or artificial sources is associated with skin cancer development and photoaging. In addition, it alters the immune system's ability to function both at the level of the skin, and in deeper tissues as well. While we can utilize UV radiation for therapeutic benefit in phototherapy for skin diseases, the risk-benefit ratio is carefully balanced for the best health outcome of the individual. The same conclusion cannot be supported for indoor tanning in which chronic UV overexposure provides no tangible health benefit with undue risk of UV radiation's deleterious effects.

References

1. George WM, Burks JW Jr (1955) Treatment of vitiligo with psoralen derivatives. AMA Arch Dermatol 71(1):14–18
2. Cox AM (1995) Porphyria and vampirism: another myth in the making. Postgrad Med J 71(841):643–644
3. Ibrahim SF, Brown MD (2008) Tanning and cutaneous malignancy. Dermatol Surg 34(4):460–474
4. Pfeifer GP, You YH, Besaratinia A (2005) Mutations induced by ultraviolet light. Mutat Res 571(1–2):19–31
5. National Cancer Institute, US National Institutes of Health (2010) SEER Stat Fact Sheet: melanoma of the skin. http://seer.cancer.gov/statfacts/html/melan.html. Accessed 11 Oct 2010
6. Leachman SA, Carucci J, Kohlmann W, 4 et al (2009) Selection criteria for genetic assessment of patients with familial melanoma. J Am Acad Dermatol 61:677 e1–677 e14
7. Robinson J (2010) Managing melanoma in the modern millennium. Annual meeting of the American Academy of Dermatology 2010. Miami, FL
8. Peae SC, Sober AJ, Tsao H et al (2008) Cutaneous melanoma. In: Wolff K, Goldsmith LA, Katz SI, Gilchrest BA, Paller AS et al (eds.) Fitzpatrick's dermatology in general medicine. McGraw Hill, New York
9. Melanoma Center. Risk factors: moles. http://www.melanomacenter.org/risk/moles.html. Accessed June 2010
10. Marrett LD, King WD, Walter SD et al (1992) Use of host factors to identify people at high risk for cutaneous malignant melanoma. Can Med Assoc J 147(4):445–453
11. Mack TM, Floderus B (1991) Malignant melanoma risk by nativity, place of residence at diagnosis, and age at migration. Cancer Causes Control 2(6):401–411
12. MacLennan R, Green AC, McLeod GR et al (1992) Increasing incidence of cutaneous melanoma in Queensland, Australia. J Natl Cancer Inst 84(18):1427–1432
13. Gilchrest BA, Eller MS, Geller AC et al (1999) The pathogenesis of melanoma induced by ultraviolet radiation. N Engl J Med 340(17):1341–1348
14. Osterlind A, Tucker MA, Stone BJ et al (1988) The Danish case-control study of cutaneous malignant melanoma. II. Importance of UV-light exposure. Int J Cancer 42(3):319–324
15. Swerdlow AJ, Weinstock MA (1998) Do tanning lamps cause melanoma? An epidemiologic assessment. J Am Acad Dermatol 38(1):89–98
16. International Agency for Research on Cancer Working Group on artificial ultraviolet (UV) light and skin cancer (2006) The association of use of sunbeds with cutaneous malignant melanoma and other skin cancers: a systematic review. Int J Cancer 120:1116–1120

17. El Ghissassi F, Baan R, Straif K et al (2009) A review of human carcinogens – part D: radiation. Lancet Oncol 10(8):751–752
18. US Food and Drug Administration (2010) Executive summary: the potential risks and benefits of tanning lamps. General and plastic surgery devices panel meeting. http://www.fda.gov/downloads/AdvisoryCommittees/CommitteesMeetingMaterials/MedicalDevices/MedicalDevicesAdvisoryCommittee/GeneralandPlasticSurgeryDevicesPanel/UCM205687.pdf. Accessed 30 July 2010
19. American Academy of Dermatology www.aad.org/gov/congressional/tanning.html. Accessed 11 Oct 2010
20. Snowden RV (2010) Study links tanning bed use to increased risk of melanoma. American Cancer Society. http://www.cancer.org/Cancer/news/News/study-links-tanning-bed-use-to-increased-risk-of-melanoma. Accessed 11 Oct 2010
21. American Academy of Dermatology. Basal cell carcinoma. http://www.aad.org/public/publications/pamphlets/sun_basal.html. Accessed June 2010
22. Jennings L, Schmults CD (2010) Management of high-risk cutaneous squamous cell carcinoma. J Clin Aesthet Dermatol 3(4):39–48
23. Leiter U, Garbe C (2008) Epidemiology of melanoma and nonmelanoma skin cancer – the role of sunlight. Adv Exp Med Biol 624:89–103
24. Situm M, Buljan M, Bulat V et al (2008) The role of UV radiation in the development of basal cell carcinoma. Coll Antropol 32(Suppl 2):167–170
25. Strickland PT, Vitasa BC, West SK et al (1989) Quantitative carcinogenesis in man: solar ultraviolet B dose dependence of skin cancer in Maryland watermen. J Natl Cancer Inst 81(24):1910–1913
26. Rodust PM, Stockfleth E, Ulrich C et al (2009) UV-induced squamous cell carcinoma – a role for antiapoptotic signalling pathways. Br J Dermatol 161(Suppl 3):107–115
27. de Gruijl FR, Rebel H (2008) Early events in UV carcinogenesis – DNA damage, target cells and mutant p53 foci. Photochem Photobiol 84(2):382–387
28. Karagas MR, Stannard VA, Mott LA et al (2002) Use of tanning devices and risk of basal cell and squamous cell skin cancers. J Natl Cancer Inst 94(3):224–226
29. Kligman LH, Kligman AM (1986) The nature of photoaging: its prevention and repair. Photodermatol 3(4):215–227
30. Urbach F, Forbes PD, Davies RE et al (1976) Cutaneous photobiology: past, present and future. J Invest Dermatol 67(1):209–224
31. Helfrich YR, Sachs DL, Voorhees JJ (2008) Overview of skin aging and photoaging. Dermatol Nurs 20(3):177–183, quiz 184
32. Bernstein EF, Chen YQ, Kopp JB et al (1996) Long-term sun exposure alters the collagen of the papillary dermis. Comparison of sun-protected and photoaged skin by northern analysis, immunohistochemical staining, and confocal laser scanning microscopy. J Am Acad Dermatol 34(2 Pt 1):209–218
33. Sellheyer K (2003) Pathogenesis of solar elastosis: synthesis or degradation? J Cutan Pathol 30(2):123–127
34. Lavker RM, Gerberick GF, Veres D et al (1995) Cumulative effects from repeated exposures to suberythemal doses of UVB and UVA in human skin. J Am Acad Dermatol 32(1):53–62
35. Gilchrest BA, Blog FB, Szabo G (1979) Effects of aging and chronic sun exposure on melanocytes in human skin. J Invest Dermatol 73(2):141–143
36. Rhodes AR, Albert LS, Barnhill RL et al (1991) Sun-induced freckles in children and young adults. A correlation of clinical and histopathologic features. Cancer 67(7):1990–2001
37. Bastiaens M, Hoefnagel J, Westendorp R et al (2004) Solar lentigines are strongly related to sun exposure in contrast to ephelides. Pigment Cell Res 17(3):225–229
38. Yaar M, Gilchrest BA (2001) Ageing and photoageing of keratinocytes and melanocytes. Clin Exp Dermatol 26(7):583–591
39. Kripke ML (1974) Antigenicity of murine skin tumors induced by ultraviolet light. J Natl Cancer Inst 53(5):1333–1336

40. Hanneman KK, Cooper KD, Baron ED (2006) Ultraviolet immunosuppression: mechanisms and consequences. Dermatol Clin 24(1):19–25
41. Tuteja N, Tuteja R (2001) Unraveling DNA repair in human: molecular mechanisms and consequences of repair defect. Crit Rev Biochem Mol Biol 36(3):261–290
42. Walterscheid JP, Nghiem DX, Kazimi N et al (2006) Cis-urocanic acid, a sunlight-induced immunosuppressive factor, activates immune suppression via the 5-HT2A receptor. Proc Natl Acad Sci USA 103(46):17420–17425
43. Aubin F (2003) Mechanisms involved in ultraviolet light-induced immunosuppression. Eur J Dermatol 13(6):515–523
44. Kang K, Gilliam AC, Chen G et al (1998) In human skin, UVB initiates early induction of IL-10 over IL-12 preferentially in the expanding dermal monocytic/macrophagic population. J Invest Dermatol 111(1):31–38
45. Teunissen MB, Piskin G, di Nuzzo S et al (2002) Ultraviolet B radiation induces a transient appearance of IL-4+ neutrophils, which support the development of Th2 responses. J Immunol 168(8):3732–3739
46. Larcom LL, Morris TE, Smith ME (1991) Tanning salon exposure suppression of DNA repair capacity and mitogen-induced DNA synthesis. Photochem Photobiol 53(4):511–516
47. Bandow GD, Koo JY (2004) Narrow-band ultraviolet B radiation: a review of the current literature. Int J Dermatol 43(8):555–561
48. Kroft EB, Berkhof NJ, van de Kerkhof PC et al (2008) Ultraviolet A phototherapy for sclerotic skin diseases: a systematic review. J Am Acad Dermatol 59(6):1017–1030
49. Honigsmann H, Szeimies RM, Knobler R et al (2008) Photochemotherapy and photodynamic therapy. In: Wolf K, Goldsmith L, Katz S, Gilchrest BA, Paller A (eds.) Fitzpatrick's dermatology in general medicine, 7th edn. McGraw Hill, New York, pp 2249–2262
50. Naldi L (2010) Malignancy concerns with psoriasis treatments using phototherapy, methotrexate, cyclosporin, and biologics: facts and controversies. Clin Dermatol 28(1):88–92
51. Rao J, Bissonnette R, Taylor C (2010 January) Photodynamic therapy for the dermatologist. eMedicine
52. Menter A, Korman NJ, Elmets CA et al (2010) Guidelines of care for the management of psoriasis and psoriatic arthritis: section 5. Guidelines of care for the treatment of psoriasis with phototherapy and photochemotherapy. J Am Acad Dermatol 62(1):114–135

Chapter 7
Tanning Dependence: Is Tanning an Addiction?

Avnee Shah, Samantha Smith, Carolyn J. Heckman, and Steven R. Feldman

Abstract Despite the association between skin cancer and ultraviolet radiation, the public continues to increase its exposure to indoor tanning. Efforts to discourage tanning have focused on the risks of skin cancer and the negative effects on appearance, and research on the motivations for indoor tanning have focused primarily on appearance-related issues. However, a growing body of observational and experimental trials now supports a physiologic mechanism contributing to high-risk tanning behavior and the existence of an addictive quality to tanning. Plausible physiologic mechanisms mediated by endorphins have been proposed. Controlled, blinded studies have conclusively demonstrated that there are physiologic effects of tanning that contribute to tanning behavior. Survey studies have provided supportive evidence of addiction-like qualities of tanning behavior. With this chapter, we explore characteristics of addiction and discuss psychological and physiological motivations to tan that go beyond appearance in order to better understand how addiction is defined and whether tanning behavior meets the definition of an addictive behavior. Future research and programs to

---

A. Shah • S. Smith
Center for Dermatology Research, The Center for Dermatology Research is supported by an unrestricted educational grant from Galderma Laboratories, L.P., Department of Dermatology, Wake Forest University School of Medicine, Winston-Salem, NC, USA

C.J. Heckman
Cancer Prevention and Control Program, Fox Chase Cancer Center, 333 Cottman Avenue, Philadelphia, PA 19111, USA
e-mail: carolyn.heckman@fccc.edu

S.R. Feldman (✉)
Departments of Dermatology, Pathology & Public Health Sciences, Wake Forest University School of Medicine, Medical Center Boulevard, Winston-Salem, NC 27157-1071, USA
e-mail: sfeldman@wfubmc.edu

C.J. Heckman and S.L. Manne (eds.), *Shedding Light on Indoor Tanning*,
DOI 10.1007/978-94-007-2048-0_7, © Springer Science+Business Media B.V. 2012

reduce excessive tanning and other ultraviolet radiation exposure will likely need to address physiologic drivers of tanning behavior and not just appearance motives for tanning.

Keywords Tanning • Dependence • Addiction • Substance-related disorder • Ultraviolet radiation • DSM-IV-TR model • Social learning and cognitive model • Behavioral model • Pharmacologic model • Withdrawal • Tolerance • Reinforcement • Endorphins • Opioids • Reward • Self-efficacy • Coping skills • Environmental stimuli

Abbreviations

ACTH	Adrenocorticotropic hormone
CAGE	Cut down Annoyed, Guilty, Eye-opener from CAGE alcoholism screening questionnaire
DSM-IV-TR	Diagnostic and Statistical Manual of Mental Disorders – 4th edition – Text revision
POMC	Pro-opiomelanocortin
US	United States
UV	Ultraviolet
αMSH	Alpha-melanocyte stimulating hormone
βLPH	Beta-lipotropic hormone

7.1 Introduction

Many people perceive "benefits" to a tan appearance. For decades, the popularity of indoor tanning activities has primarily been attributed to cosmetic motivations. The mainstream US cultural belief that tanned skin implies wealth and beauty has become well-established and is evidenced by magazine ads and television programs displaying tanned celebrities. One study in the literature found that 81% of young adults indeed believe that tans improve appearance [1]. Come prom time for high-school girls, tanning becomes a very popular pursuit. Studies have shown that the prevalence of indoor tanning behavior in white females in the US more than doubles from 15% at age 15 to 35% at age 17 [2].

Though there is a clear association between UV exposure and the development of skin cancer, indoor and outdoor exposure to UV continues to increase [3, 4]. Even more intriguing is the finding that tanners have been reported to be well-educated and knowledgeable about the dangers of using tanning lamps, yet still choose to tan and not alter this behavior in light of their awareness [5].

Aesthetic motivation is an important driver of tanning behavior, but the individuals who are at the greatest risk of problems from tanning engage in tanning behavior far more often than is necessary to maintain a tanned appearance. Some of the most frequent tanners tan to the point at which their skin becomes leathery and mottled, raising serious questions about whether they are tanning primarily for appearance-based reasons.

These observations suggest that there are strong forces behind the development of tanning behavior, which may involve more than just aesthetic pursuits. Several studies regarding tanning have proposed that UV exposure has addictive properties similar to other abused substances. Research into the existence and potential mechanisms of tanning dependence provides biological evidence for this phenomenon. For example, UV exposure may have the ability to up-regulate genes in the skin, leading to the release of endorphins, a type of natural opioid that is involved in the brain's reward pathway [6]. The production of feel-good endorphins in some tanning bed users is an incentive to return for more tanning to reproduce this endorphin "high." Much of the focus of studies and interventions on tanning behavior has concentrated on appearance-related issues. With this chapter, however, we hope to (1) aid in the explanation of what is described as "tanning dependence," (2) help address the question, "Is excessive tanning an addiction?", and (3) provide the reader with an understanding that the physiologic effects resulting from UV exposure may also contribute to the behavior of tanning.

7.2 Defining Substance Abuse and Dependence

The Diagnostic and Statistical Manual of Mental Disorders – Fourth edition – Text revision (DSM-IV-TR) describes substance abuse as a pattern of repeated use of a substance, which leads to adverse consequences in major life areas such as occupational, physical or social functioning [7]. Substance dependence, commonly referred to as addiction, refers to a more significant impairment caused by escalating use of the substance despite its negative physical and/or psychological effects. A diagnosis of substance dependence requires that three or more of the following seven criteria are met within a 12-month period [7].

Criterion 1 for substance dependence is tolerance. Tolerance refers to the phenomenon that develops after repetitive use of a substance in which there is a need for increased amounts of the substance over time to achieve the desired effect. Tolerance can also refer to a decreased effect obtained with continued usage of the same amount of a substance.

Criterion 2 is withdrawal. Withdrawal refers to the unpleasant symptoms that occur upon discontinuation or decrease in dosage of the substance after prolonged usage. Withdrawal symptoms vary depending on the class of substance used. For example, symptoms of withdrawal from narcotics include nausea, vomiting, sweating, anxiety, and craving for the substance. Whereas, withdrawal from cocaine causes depression, fatigue, paranoia, irritability, hunger, as well as craving for more cocaine. Generally, the symptoms of withdrawal are usually opposite that of the acute effects produced by the substance. A hallmark of dependence is repeated or continued use of the substance in order to avoid the symptoms of withdrawal.

The rest of the criteria (criteria 3–7) all surround the compulsiveness of substance-related behavior. Criterion 3 involves difficulty controlling use of the substance, exemplified by individuals' exposing themselves for longer times or to larger amounts of the substance than initially intended. Criterion 4 refers to the behavior of trying to cut down on use of the substance. In an individual's history, multiple unsuccessful attempts at discontinuation or decrease in dosage of the substance may be found. Criterion 5 refers to increased time spent in obtaining and consuming the substance. Criterion 6 points to the social, occupational, or recreational activity involvement that is reduced or given up due to substance use, i.e., spending less time with family and in hobbies and more time using the substance. Lastly, criterion 7 is the persistent usage of the substance despite experiencing its adverse effects (as might be seen in a patient who has developed skin cancer but who continues to tan).

Increasingly, it has been suggested that certain activities and behaviors, such as gambling, shopping, internet use, etc., can incite addiction similar to that of chemical substances. "Tanning dependence" can be viewed as a "substance"-related disorder in which ultraviolet radiation functions as the substance. For example, the aforementioned criteria concerning difficulty controlling usage and continued usage despite negative consequences are demonstrated in the many tanners who tan significantly beyond the amount needed to become tan, and continue their behavior despite experiencing negative consequences like sunburn and skin cancer. Tolerance develops as higher doses of UV are required over time in order for the skin to darken in response to UV exposure; whether tolerance develops to the feel good effect has not been explored. There is, however, evidence for physiologic withdrawal symptoms such as disorientation, difficulty concentrating, fatigue, and nausea [8].

7.3 Psychosocial Evidence for Tanning Dependence

Multiple survey studies have explored the concept of dependence in relation to tanning behavior using the DSM-IV-TR criteria for substance-related disorders. See Table 7.1 for a summary of major studies reviewed in this chapter. The first study, by Warthan and colleagues [9], assessed 145 beach-going participants via a modified CAGE (*C*ut Down, *A*nnoyed, *G*uilty, *E*ye-Opener) Questionnaire, usually employed to screen for alcohol problems, in addition to a modified DSM-IV-TR substance dependence criteria employed by the American Psychiatric Association. The CAGE questionnaire involves answering the following four questions, with two or more "yes" responses suggesting a likely tanning problem: Cut Down – "Do you try to cut down the time you spend in the sun, but find yourself still suntanning?"; Annoyed – "Do you ever get annoyed when people tell you not to sun tan?"; Guilty – "Do you ever feel guilty that you are in the sun too much?"; Eye-Opener – "When you wake

Table 7.1 Major studies summary table

Study	Description	Key findings and conclusions
Knight et al. [5]	Multiple choice questionnaire designed to help correlate frequency of indoor tanning with knowledge of risks and benefits of UV exposure.	Over 90% of indoor tanners were aware of the risks associated with the behavior.
Warthan et al. [9]	Modified CAGE and DSM-IV criteria questionnaires were used to assess the presence of a UV substance-related disorder.	Of the study participants, 26% met modified CAGE criteria, and 53% met modified DSM-IV criteria, for UV substance-related disorder. Frequently tanning individuals may have a UV substance-related disorder.
Poorsattar et al. [10]	A modified CAGE questionnaire was distributed among undergraduate college students to determine the prevalence of UV substance-related disorders.	28% of indoor tanning students in the study tested positive on the modified CAGE questionnaire for UV substance-related disorder. Positive results were more likely if the student was a woman, indoor tanner, or frequent tanner.
Harrington et al. [12]	A modified CAGE and DSM-IV-TR questionnaire given to tanning salon patrons.	41% met "tanning addictive disorder" criteria. 33% met criteria for problematic tanning behavior.
Zeller et al. [13]	Telephone interviews conducted to explore whether indoor tanning leads to dependency.	Individuals starting indoor tanning at a younger age as well as individuals tanning with a higher frequency reported more difficulty quitting indoor tanning.
Feldman et al. [14]	Controlled, blinded, repeated choice trial in which frequently tanning individuals were exposed to identical UV and non-UV tanning beds twice weekly for 6 weeks and then given the option to tan in one bed a third time. The primary outcome was the percentage of choice sessions of the UV bed.	95% of choice sessions were for the UV bed. Frequent tanners under blinded conditions can distinguish UV stimuli from non-UV stimuli. UV is a reinforcing stimulus.
Wintzen et al. [19]	An investigation into expression of POMC and its derivatives by normal human keratinocytes.	Keratinocytes have the ability to synthesize POMC and its derivatives, a process that can be influenced by UV. Melanogenesis in the skin after UV exposure may involve the POMC derivatives βLPH and β-endorphin. These molecules released into circulation may have systemic effects.

(continued)

Table 7.1 (continued)

Study	Description	Key findings and conclusions
Taylor et al. [24]	Controlled study conducted to determine ability of UV to alleviate pain in patients with Fibromyalgia. Patients received both UV and non-UV stimuli for 2 weeks followed with a 6-week randomization to UV or non-UV stimuli and asked to rate pain severity.	During initial 2 weeks, UV exposure significantly decreased pain scores to a greater degree than non-UV stimuli. After 6-week randomization period, there was mild improvement in pain for UV exposure group versus non-UV exposure group. Tanning beds may have potential to reduce pain in patients with Fibromyalgia.
Schauer et al. [31]	A study of human normal keratinocytes and an epidermal carcinoma cell line.	Human keratinocytes produced POMC-derived peptides. Production of alpha MSH and ACTH were up-regulated upon treatment with UV.
Kaur et al. [23]	Double-blind, placebo-controlled trial of UV and non-UV stimuli in frequent and infrequent tanners combined with placebo-controlled opioid blockade using naltrexone.	Using opioid blockage, frequent tanners' preference for UV stimuli was reduced. Withdrawal-like symptoms were exhibited by half of the frequent tanners tested upon administration of opioid blocker.
Kaur et al. [33]	Double-blind, placebo-controlled, randomized trial conducted to evaluate plasma β-endorphin levels in frequent and infrequent tanners in response to exposure to UV.	No trend in plasma β-endorphin levels seen.
Mosher and Danoff-Burg [17, 18]	A modified CAGE and DSM-IV-TR questionnaire given to college students.	39.3% met DSM-IV-TR criteria and 30.6% met CAGE criteria. Tanning dependent students had greater anxiety and other substance use. Indoor tanning was associated with anxiety disorders among men and other substance use among women.

This table provides descriptions and key findings of the major studies mentioned in this chapter

up in the morning do you want to sun tan?". The DSM-IV-TR-based questionnaire is considered positive for tanning dependence if three of the following seven symptoms that typically describe substance dependence occurred in association with the participant's tanning activity in the previous 12 months: tolerance, withdrawal, more use over a longer period than intended, unsuccessful efforts to cut down, increased time spent acquiring the substance (in this case, UV radiation), social/occupational/ recreational activity involvement reduced or abandoned, and continued use despite adverse effects. Warthan and colleagues found that 26% of the study population met modified CAGE criteria for a UV substance-related disorder, while 53% met modified DSM-IV-TR criteria for a UV substance-related disorder, lending support to the idea of "tanning dependence" as well as a reinforcing stimulus to tan [9]. Although the "hit rate" on the two questionnaires is discordant, the findings suggest that several established features of addiction can be found in a subset of the tanning population.

A similar study by Poorsattar and Hornung was conducted using a modified CAGE questionnaire to determine if symptoms of a substance related disorder due to UV were present in 385 college students [10]. Twenty-eight percent of the indoor tanners in the study tested positive to the modified CAGE questionnaire, whereas only 12% of the non-indoor tanners tested positive. Forty-four percent of the indoor tanners also reported a feeling of relaxation as their motivation for using UV beds. These authors, along with those of similar studies, have concluded that frequent tanning represents a type of substance-related disorder that is salient to and common among college-aged individuals [10, 11]. A recent study of addictive-like behaviors in indoor tanning salon users by Harrington and colleagues surveyed 100 subjects (n=100) using a modified CAGE questionnaire and modified DSM-IV substance dependence criteria. Forty-one percent of participants were found to meet "tanning addictive disorder" criteria, and an additional 33% met criteria for problematic tanning behavior [12].

One study focused specifically on perceived difficulty controlling indoor tanning behavior. This was examined by telephone interviews of 1,275 adolescents in the Minneapolis-St. Paul and Boston areas [13]. Participants were asked "How hard would it be for you to stop tanning indoors?" with scores of 0–3 indicating not at all and 7–10 indicating extremely. The study found that 20.9% of its participants reported that they would have difficulty quitting tanning. An earlier age of tanning initiation and a higher frequency of tanning bed usage were found among those reporting that they would have difficulty quitting indoor tanning. The presence of a perceived loss of control in relation to tanning bed usage further supports an expanding profile of dependence in the adolescent and young adult indoor tanning population [13].

Several studies have investigated the reinforcing effects of indoor tanning. For example, 14 frequent-tanning (defined as tanning 8–15 times per month) young adult patients spent time in both UV and non-UV tanning beds as part of a controlled, single-blinded study to determine if UV causes a response that is not produced by the non-UV bed. For 6 weeks, the subjects were exposed to two tanning beds on both Monday and Wednesday; the beds were identical except for the exposure to UV light in one bed and the lack of exposure to UV in the other. The participants were unaware of which was the UV bed, but were given the option to tan again in the bed of their choosing on the Friday of each week. The participants' mood before and after tanning in each bed was also assessed by having participants rate ten different mood states on a scale from 1 to 9 [14]. At 95% of the open Friday sessions, subjects chose to tan in the UV bed. Greater relaxation scores and lower tension scores were reported after UV exposure versus non-UV exposure, and upon questioning, the strongest motivation for the bed choice among these patients was feelings of relaxation [14]. Interestingly, one subject suffered from chronic lower back pain and reported amelioration of pain to a greater degree when tanning in the UV bed as opposed to the non-UV bed [15].

The positive effects – mood enhancement and pain relief – that frequent tanners gain from UV tanning exposure serve as reinforcing stimuli for the behavior. The warm, quiet, relaxing environment of a tanning bed may contribute to behavioral reinforcement, but in addition, the effects are due, at least in part, specifically to UV light and not just a warm, relaxing tanning experience. Although the study was blinded and the participants did not know the UV status of the beds because they both produced similarly perceived light and heat, the participants overwhelmingly favored the UV bed over the non-UV bed. This reinforcement associated with UV exposure leads to habitual tanning as a kind of conditioned response, or a behavioral learning of tanning addiction.

Mood enhancement is an interesting benefit of UV exposure that can be a significant incentive for tanning [16]. Effects on mood and the mental state of the user may be related to the development of dependence in some tanners. Researchers of tanning addiction found a higher prevalence of anxiety and other substance use in subjects who were tanning-dependent based on DSM-IV and CAGE questionnaires [17, 18]. Anxiety and depression frequently occur comorbidly with substance dependence as well, suggesting an association between those psychiatric disorders and a tendency towards addiction. This connection also represents an opportunity for much-needed intervention in tanning dependence. Screening for and treating underlying mood disorders in frequent tanners may help to curb their behavior and reduce their future skin cancer risk.

7.4 Physiologic Evidence for Tanning Dependence

Physical dependence occurs when repetitive use of a substance causes alterations in the body's homeostatic set points and drug-receptor interactions. In the presence of large amounts of a substance, receptors in the nervous system are up-regulated so that when the drug is eliminated, the normal state is perceived as deficient and the system can't function properly. Tolerance and withdrawal (refer to Sect. 7.3 DSM-IV-TR) are hallmarks of this physiologic adaptation in that the body produces a physical response to increased use and then discontinuation. Most addictive substances activate dopaminergic reward pathways and trigger endorphin release, which produces a euphoric feeling in the user. The effects can be very rapid and pronounced, which strengthens the association with the drug and leads to reinforcement of the drug-use behavior. The reward system in tanning is thought to occur through endorphins as well, as some studies have demonstrated the release of endorphin opioid compounds by melanocytes after in vitro exposure to UV radiation [19]. These opioids potentially could have local or central effects, including the possibility of increasing dopamine release in the pleasure center of the brain, launching a process of reinforcement the same way that other addictive substances would. [20].

A small, randomized controlled trial conducted to see if opioid blockade could prevent frequent tanners from distinguishing UV versus non-UV stimuli yielded interesting results. In this study, three female, frequent tanners (defined as tanning between 8 and 15 times per month) were treated with 50 mg of naltrexone before UV and non-UV exposure [8]. Naltrexone blocks opioid receptors, may precipitate withdrawal in those with high opioid levels in their system, but is well-tolerated by normal subjects [21, 22]. Within the study, all three subjects preferred the non-UV bed, indicating that by blocking opioid receptors, naltrexone seemed to decrease the frequent tanners' preference for UV that had been seen in previous studies [14]. Interestingly, two subjects had unexpected adverse effects from receiving the naltrexone. One participant felt "disoriented" and "unable to concentrate" post naltrexone administration and felt nauseous later in the day. The second participant felt "emotionally drained" shortly after administration of the medication and developed nausea and vomiting as well [8]. This suggests that frequent tanning may cause chronically elevated levels of opioids, since some subjects experienced withdrawal-like symptoms when given the opioid-blocking naltrexone.

The authors expanded upon those findings by trying to induce withdrawal symptoms in frequent tanners by blocking the opioid-like effects of UV exposure. In this randomized, double-blind, placebo-controlled trial, eight frequent tanners and eight infrequent tanners were exposed to UV and non-UV stimuli in combination with placebo-controlled administration of 5, 15, and 25 mg doses of naltrexone [23]. The participants were given the medication and then exposed to UV and non-UV tanning beds and asked about which they preferred. When given the placebo or the 5 mg naltrexone dose, all frequent tanners favored the UV bed. However, when given the 15 and 25 mg naltrexone doses, the frequent tanners' preference for the UV bed diminished. Adverse events were not experienced by any subject in the infrequent tanning group at any medication doses, however, on administration of the 15 mg naltrexone dosage, four of the eight frequent tanners experienced nausea and/or jitteriness, causing two to withdraw from the trial [23]. The adverse events witnessed in this study are consistent with withdrawal symptoms seen in opioid-dependant patients treated with an opioid antagonist, further lending support to the notion of an addictive effect of UV in a pharmacologic model of tanning dependence that involves opioids in the reinforcing effects of UV.

Based on these findings suggesting a physiologic effect mediated by endorphins and on anecdotal reports that tanning ameliorated back pain, a controlled study was conducted to determine if UV could be used to alleviate pain in patients with fibromyalgia [24]. Nineteen subjects with fibromyalgia were enrolled in a trial in which they received both UV and non-UV stimuli for 2 weeks followed with a 6-week randomization to UV or non-UV stimuli. They were asked to rate pain severity on a variety of scales including an 11-point Likert scale and The McGill Pain Questionnaire [25]. During the first 2 weeks, pain scores were reported to have decreased 0.44 points out of 11 after UV exposure from baseline. UV exposure reduced pain levels to a greater degree than non-UV exposure ($p=0.0059$). After the 6-week randomization period, there was a mild improvement in pain for the UV exposure group in comparison to the non-UV exposure group ($p=0.049$). Four weeks after the last treatment, pain scores showed no significant difference between the two groups [24]. The results from this pilot study of pain reduction in patients with fibromyalgia further supports the likelihood that physiologic effects of UV contribute to frequent tanners' tanning behavior.

Finally, frequent tanner's preference for a UV over a non-UV bed was confirmed in a recent study by Kourosh and colleagues. In addition, UV exposure increased regional cerebral blood flow compared to sham light as assessed by single photon emission tomography [26]. This finding is similar to what has been found during exposure to recreational drugs.

7.5 Proposed Mechanisms Underlying Tanning Dependence

All of this research supports the existence of tanning dependence as a substance-related disorder but does not fully explain how UV exposure leads to this dependence. With dependence to substances such as alcohol and cocaine, there is a neurologic reward pathway that is activated by the drug, causing the subsequent release of psychoactive chemicals including β-endorphin, an endogenous opioid known to have analgesic and euphoric effects [27]. The aforementioned studies using naltrexone suggest that opioids underlie tanning dependence as well, but how UV may lead to opioid release is unclear.

Researchers at Boston University studying the mechanisms underlying pigmentation made a discovery that supports the idea of a UV-triggered reward process. Their research involved the compound proopiomelanocortin (POMC). This large protein is cleaved into several other biologically important substances such as alpha-melanocyte stimulating hormone (αMSH), which stimulates melanin pigment secretion in the skin [28], and β-endorphin, which acts on opioid receptors throughout the body. Endorphins have analgesic properties, can induce euphoria, and have roles in behavior and within the reward and reinforcement pathways involved in substance dependence [27].

It was previously believed that POMC was produced mainly in the pituitary gland, and the derivatives that were cleaved from it traveled to other parts of the body like the skin, where αMSH has been found in abundance [29, 30]. However, scientists at the University of Munster and Oregon Health and Sciences University discovered that some of the αMSH in skin had not come from cleaved POMC in the pituitary, but rather from the skin itself [31, 32]. These studies not only reported for the first time that normal human keratinocytes in the skin produce their own POMC derivatives but also suggested that UVB irradiation of keratinocytes has the potential to increase the synthesis and secretion of these substances [31].

This hypothesis was tested in a study which examined the effect of UV stimulation on keratinocyte levels of POMC and two of its derivatives, β-lipotropic hormone (βLPH) and β-endorphin. Keratinocyte cultures were exposed to 10–30 mJ/cm^2 of UV stimuli, an amount found within the physiologic range of radiation that is able to reach the basal layer of the epidermis (outer layer of the skin), and then levels of POMC and its derivative compounds were measured using standard laboratory techniques [19]. The keratinocytes produced some POMC, βLPH, and β-endorphin at baseline before UV stimulation, with significant increases in the amounts after UV exposure [19]. This suggests that a UV-induced endorphin surge could underlie the reward system associated with tanning behavior and can be likened to the endorphin release seen after exercising that is responsible for the euphoric state described as a "runner's high."

The evidence for UV's ability to induce increased levels of serum endorphin is somewhat conflicting, however. While both the *in vitro* and *in vivo* release of endorphins after UV exposure have been documented, other studies failed to corroborate this phenomenon [19, 33–37]. For example, a more recent report describes a small, double-blinded, placebo-controlled, randomized trial conducted to assess levels of β-endorphins in plasma before and after UV and non-UV exposure. Blood was drawn from three frequent and three non-frequent tanners at baseline and after both UV and non-UV exposure. The study measured β-endorphin levels on the same day samples were obtained and did not detect an increase in levels after exposure to UV; however, the study sample size was small [33]. Much more research is needed to fully clarify the mechanism of tanning dependence and the exact role that endorphins play.

Another theory behind UV's physiologic effect on inducing dependence surrounds the pineal melatonin-generating system, and more specifically, the enzyme N-acetyltransferase. N-acetyltransferase functions to convert serotonin to melatonin. UV inhibits this enzyme, leading researchers to postulate that indoor tanners may be able to use their exposure to increase their levels of serotonin. Given serotonin's established role in the pleasure pathways of the brain, this may represent another physiological mechanism underlying tanning dependence [13], but it has not been well-established in the literature.

7.6 Future Research

Our knowledge of tanning dependence is still in its infancy, and there is potential for future research and developments in the field, such as cue response, brain imaging, and interventional investigations. Studies in which physiologic parameters associated with craving are measured in response to tanning paraphernalia would be analogous to studies already performed with other substances and could further our understanding of tanning-related disorders. The studies of nicotine and crack cocaine addiction have reported increased cravings and increased heart rates during exposure to drug-related stimuli in dependent individuals [38, 39]. By exposing subjects to different levels of tanning stimuli, ranging from conversation with keywords such as "tanning" and "sun" to the scents of tanning lotions and sights of tanning beds, these studies have the ability to measure whether similar cravings and cue responses exist in tanning dependence as compared to other drugs.

Brain imaging is another area that holds promise for future research into tanning dependence. Functional magnetic resonance images (MRIs) have been used to measure changes in brain activity in response to drug-related stimuli. Studies have shown increased activation of the brain's pleasure centers, such as the nucleus accumbens, in response to substance-related stimuli in smokers as well as cocaine addicts [40, 41]. These same techniques could help uncover the biology of tanning dependence, and what happens in the brains of frequent tanners. Such work is beginning to appear [26].

The elucidation of tanning dependence may play a large role for future efforts to reduce tanning behavior, particularly in the frequent tanner. Promotion and acceptance of natural skin color as a societal standard is crucial, but this type of change may not address the physiologic drivers of tanning that are at work among frequent tanners. More intensive individual approaches that target addiction and dependence such as medication and therapy, may need to be developed to address these problems.

7.7 Conclusion

The development of tanning behavior is most likely the result of a combination of biological, psychological and social factors, rather than one single element. In addictive processes like alcohol and nicotine dependence, there are biopsychosocial interactions at play, which is likely true for tanning dependence as well. What may start as a cosmetic or socially-motivated process could become physiologically reinforced in some people. Those tanners who meet DSM-IV-TR criteria for dependence are much more likely to abuse other substances like alcohol and cigarettes, suggesting that there are similar processes underlying addiction to both chemical substances and ultraviolet radiation.

Despite its documented risks, many people perceive that tanning provides them numerous benefits. Improved appearance is the most common one cited, but tanners in various studies have also specified enhanced mood, tanning in advance of vacation to prevent sunburn, improved energy, increased vitamin D levels, and relaxation as reasons to use indoor tanning beds [16]. Cancer risks notwithstanding, tanning bed use does not necessarily indicate a substance dependence-like disorder. However, when tanning is practiced beyond the amount needed to maintain a tan and the tanner feels unable to stop, it qualifies as an addiction by most definitions of dependence. A subset of tanners meets this qualification. The nature of how tanning becomes an addiction has yet to be fully elucidated, but the findings of central striatal activation strongly supports a central mechanism [26].

References

1. Robinson JK, Kim J, Rosenbaum S et al (2008) Indoor tanning knowledge, attitudes, and behavior among young adults from 1988–2007. Arch Dermatol 144(4):484–488
2. Tran TT, Schulman J, Fisher DE (2008) UV and pigmentation: molecular mechanisms and social controversies. Pigment Cell Melanoma Res 21(5):509–516

3. Swerdlow AJ, Weinstock MA (1998) Do tanning lamps cause melanoma? An epidemiologic assessment. J Am Acad Dermatol 38(1):89–98
4. Council on Scientific Affairs (1989) Harmful effects of ultraviolet radiation. JAMA 262(3): 380–384
5. Knight JM, Kirinich AN, Farmer ER et al (2002) Awareness of the risks of tanning lamps does not influence behavior among college students. Arch Dermatol 138(10):1311–1315
6. Cui R, Widlund HR, Feige E et al (2007) Central role of p53 in the suntan response and pathologic hyperpigmentation. Cell 128(5):853–864
7. American Psychiatric Association, Task Force on DSM-IV (2000) Diagnostic and statistical manual of mental disorders: DSM-IV-TR. American Psychiatric Association, Washington, DC
8. Kaur M, Liguori A, Fleischer AB Jr et al (2005) Side effects of naltrexone observed in frequent tanners: could frequent tanners have ultraviolet-induced high opioid levels? J Am Acad Dermatol 52(5):916
9. Warthan MM, Uchida T, Wagner RF Jr (2005) UV light tanning as a type of substance-related disorder. Arch Dermatol 141(8):963–966
10. Poorsattar SP, Hornung RL (2007) UV light abuse and high-risk tanning behavior among undergraduate college students. J Am Acad Dermatol 56(3):375–379
11. Heckman CJ, Egleston BL, Wilson DB et al (2008) A preliminary investigation of the predictors of tanning dependence. Am J Health Behav 32(5):451–464
12. Harrington CR, Beswick TC, Leitenberger J et al (2011) Addictive-like behaviours to ultraviolet light among frequent indoor tanners. Clin Exp Dermatol 36(1):33–38
13. Zeller S, Lazovich D, Forster J et al (2006) Do adolescent indoor tanners exhibit dependency? J Am Acad Dermatol 54(4):589–596
14. Feldman SR, Liguori A, Kucenic M et al (2004) Ultraviolet exposure is a reinforcing stimulus in frequent indoor tanners. J Am Acad Dermatol 51(1):45–51
15. Kaur M, Feldman SR, Liguori A et al (2005) Indoor tanning relieves pain. Photodermatol Photoimmunol Photomed 21(5):278
16. Ezzedine K, Malvy D, Mauger E et al (2008) Artificial and natural ultraviolet radiation exposure: beliefs and behaviour of 7200 French adults. J Eur Acad Dermatol Venereol 22(2):186–194
17. Mosher CE, Danoff-Burg S (2010) Addiction to indoor tanning: relation to anxiety, depression, and substance use. Arch Dermatol 146(4):412–417
18. Mosher CE, Danoff-Burg S (2010) Indoor tanning, mental health, and substance use among college students: the significance of gender. J Health Psychol 15(6):819–827
19. Wintzen M, Gilchrest BA (1996) Proopiomelanocortin, its derived peptides, and the skin. J Invest Dermatol 106(1):3–10
20. Nolan BV, Taylor SL, Liguori A et al (2009) Tanning as an addictive behavior: a literature review. Photodermatol Photoimmunol Photomed 25(1):12–19
21. Perez-Reyes M, Wall ME (1981) A comparative study of the oral, intravenous, and subcutaneous administration of 3H-naltrexone to normal male volunteers. NIDA Res Monogr 28:93–101
22. Gonzalez JP, Brogden RN (1988) Naltrexone. A review of its pharmacodynamic and pharmacokinetic properties and therapeutic efficacy in the management of opioid dependence. Drugs 35(3):192–213
23. Kaur M, Liguori A, Lang W et al (2006) Induction of withdrawal-like symptoms in a small randomized, controlled trial of opioid blockade in frequent tanners. J Am Acad Dermatol 54(4):709–711
24. Taylor SL, Kaur M, LoSicco K et al (2009) Pilot study of the effect of ultraviolet light on pain and mood in fibromyalgia syndrome. J Altern Complement Med 15(1):15–23
25. Melzack R (1987) The short-form McGill Pain Questionnaire. Pain 30(2):191–197
26. Kourosh AS, Harrington CR, Adinoff B (2010) Tanning as a behavioral addiction. Am J Drug Alcohol Abuse 36(5):284–290
27. Roth-Deri I, Green-Sadan T, Yadid G (2008) Beta-endorphin and drug-induced reward and reinforcement. Prog Neurobiol 86(1):1–21

28. Molina PE (2006) Endocrine physiology (Lange physiology series), 2nd edn. McGraw-Hill Medical, New York, pp 45–69
29. Tsong SD, Phillips D, Halmi N et al (1982) ACTH and beta-endorphin-related peptides are present in multiple sites in the reproductive tract of the male rat. Endocrinology 110(6):2204–2206
30. Thody AJ, Ridley K, Penny RJ et al (1983) MSH peptides are present in mammalian skin. Peptides 4(6):813–816
31. Schauer E, Trautinger F, Kock A et al (1994) Proopiomelanocortin-derived peptides are synthesized and released by human keratinocytes. J Clin Invest 93(5):2258–2262
32. Bateman A, Singh A, Kral T et al (1989) The immune-hypothalamic-pituitary-adrenal axis. Endocr Rev 10(1):92–112
33. Kaur M, Liguori A, Fleischer AB Jr et al (2006) Plasma beta-endorphin levels in frequent and infrequent tanners before and after ultraviolet and non-ultraviolet stimuli. J Am Acad Dermatol 54(5):919–920
34. Chakraborty AK, Funasaka Y, Slominski A et al (1996) Production and release of proopiomelanocortin (POMC) derived peptides by human melanocytes and keratinocytes in culture: regulation by ultraviolet B. Biochim Biophys Acta 1313(2):130–138
35. Levins PC, Carr DB, Fisher JE et al (1983) Plasma beta-endorphin and beta-lipoprotein response to ultraviolet radiation. Lancet 2(8342):166
36. Wintzen M, Ostijn DM, Polderman MC et al (2001) Total body exposure to ultraviolet radiation does not influence plasma levels of immunoreactive beta-endorphin in man. Photodermatol Photoimmunol Photomed 17(6):256–260
37. Gambichler T, Bader A, Vojvodic M et al (2002) Plasma levels of opioid peptides after sunbed exposures. Br J Dermatol 147(6):1207–1211
38. Carter BL, Robinson JD, Lam CY et al (2006) A psychometric evaluation of cigarette stimuli used in a cue reactivity study. Nicotine Tob Res 8(3):361–369
39. Saladin ME, Brady KT, Graap K et al (2006) A preliminary report on the use of virtual reality technology to elicit craving and cue reactivity in cocaine dependent individuals. Addict Behav 31(10):1881–1894
40. Breiter HC, Gollub RL, Weisskoff RM et al (1997) Acute effects of cocaine on human brain activity and emotion. Neuron 19(3):591–611
41. David SP, Munafo MR, Johansen-Berg H et al (2005) Ventral striatum/nucleus accumbens activation to smoking-related pictorial cues in smokers and nonsmokers: a functional magnetic resonance imaging study. Biol Psychiatry 58(6):488–494

Chapter 8
Selected Indoor Tanning Myths and Controversies

Gary Mendese and Barbara A. Gilchrest

Abstract In the face of increasing evidence that indoor tanning is harmful, tanning enthusiasts and the tanning industry defend the practice on several grounds. The principal argument offered in defense of year-round tanning is the claimed health benefit of high levels of vitamin D, also called the sunshine vitamin, which is made in skin following UV irradiation. However, vitamin D is readily available as an oral supplement; and oral vitamin D obviates the needed exposure to UV light that also leads to photoaging and skin cancer. Further, the claimed health benefits of high vitamin D levels are unproven. A second myth, that indoor tanning is safer than sun tanning ignores the fact that tanning is a well-established DNA damage response, achieved in proportion to DNA damage, regardless of the exact wavelengths emitted by tanning bulbs or by the sun. Similarly, obtaining a "base tan" before a planned sunny vacation will not decrease cumulative UV damage. Finally, indoor tanning may be occasionally an effective, more convenient and less expensive alternative to physician-supervised phototherapy for patients with UV-responsive skin disease, but for most patients professional staff and medical light sources with specific spectral output established as optimal for their disease provide far superior safety and efficacy.

Keywords Tanning beds • Tanning myths • Tanning controversies • Tanning industry • Indoor tanning • Indoor Tanning Association (ITA) • Vitamin D • Vitamin D supplements • Vitamin D deficiency • Vitamin D levels • Base tan • Ultraviolet adverse effects • DNA damage • Phototherapy • Year-round exposure

---

G. Mendese • B.A. Gilchrest (✉)
Department of Dermatology, Boston University School of Medicine,
609 Albany Street, J105, Boston, MA 02118, USA
e-mail: Gary.Mendese@bmc.org; bgilchre@bu.edu

C.J. Heckman and S.L. Manne (eds.), *Shedding Light on Indoor Tanning*,
DOI 10.1007/978-94-007-2048-0_8, © Springer Science+Business Media B.V. 2012

Abbreviations

AI	Adequate intake
FDA	Food and Drug Administration
IARC	International Agency for Research on Cancer
ITA	Indoor Tanning Association
IU	International units
PTH	Parathyroid hormone
UV	Ultraviolet
UVA	Ultraviolet A
UVB	Ultraviolet B

8.1 Introduction

Since the beginning of human civilization, beauty and its perception have been important on both personal and social levels. Those who are deemed more attractive often have greater self-confidence and enjoy greater social and professional success. To this end, people have always been interested in their appearance and have gone to great lengths to improve it by various means.

In Europe and the U.S., tanning as a fashion statement dates to the early twentieth century. For centuries, pale skin was perceived as a mark of wealth, refinement and beauty, while tanned skin was associated with poorer manual laborers [1]. Prior to the 1920s, American and European women took precautions to maintain a light skin tone – parasols and large hats were considered essential summer accessories [2]. In fact, at this time, magazines advertised powders that would conceal a tan; and Elizabeth Arden marketed Bleaching Cream, designed to bleach tanned skin [3].

All this changed when, in 1923, French designer Coco Chanel was seen leaving a nobleman's yacht with a deep tan after cruising from Paris to Cannes. While her tan was initially attributed to accidental sun exposure, she promoted it as part of a new natural look. An icon of her day, Chanel and her famous tan changed the concept of elegance and beauty, and soon pale was passé. Coincidentally, with the Industrial Revolution, the working classes had moved from the fields to the factories, converting time outdoors to an indulgence of the leisure class.

It wasn't until 1978, the year the first tanning bed was introduced in Arkansas, that out-of-season tans could be achieved via artificial means [4]. Previously, tans were restricted to the summer season or, in winter, to those wealthy enough to enjoy tropical vacations. Since that time, however, millions of Americans have been able to keep modest and even deep tans throughout the year by indoor tanning. These beds expose the user to high levels of carcinogenic and unregulated radiation and the indoor tanning industry as a whole remains controversial at a minimum.

People often engage in indoor tanning because they believe doing so helps them attain culturally valued attractiveness standards [5]. However, another potential reason for tanning via indoor tanning beds includes a potential addictive effect of UV exposure, discussed elsewhere in this book (see Chap. 7 by Shah and colleagues).

An estimated 30 million Americans use tanning beds annually including over 2 million adolescents [6]. The Indoor Tanning Association (ITA) reports that the industry employs over 160,000 people and generates $5 billion in revenue per year [7]. As evidence has emerged that UV exposure and specifically tanning bed use injure skin [6], justifications for tanning have also emerged. Still, because many people wish to tan and because indoor tanning is big business, myths (unfounded but prevalent beliefs) and controversies persist, particularly in the popular media and on the Internet. Four of these are discussed in this chapter.

8.2 Myth #1: Year-Round UV Exposure Is Essential for Good Health

The ITA often argues that health benefits justify indoor tanning [8]. Although most tanners report improved aesthetics or self-esteem as their primary reasons to tan [5], the better health rationale may assuage their guilt for indulging in this otherwise self-destructive behavior. Claims of disease prevention are the industry's best argument against legislation to restrict tanning bed use among teenagers, a major portion of their customer base [9].

The tanning industry argument is based on the only well-established benefit of UV exposure, vitamin D photosynthesis. For most people ambient UV exposure is the major source of their "sunshine vitamin." It is also well established that in the northern sections of the U.S., the UVB (290–315 nm) wavelengths that produce vitamin D (more correctly, pre-vitamin D) from cell membranes in the epidermis (outer layer of skin) are absent from sunlight for several months each year, depending on latitude and resulting angle of the sun as it traverses the Earth's atmosphere [10]. As a consequence, vitamin D levels for most people in the U.S. and other temperate climates are higher in the summer and lower in the winter. There is also evidence that the incidence or prevalence of several diseases is higher among people who have lower than average levels of vitamin D [11]. In combination, these observations have led the indoor tanning industry to argue that year-round exposure to UV is beneficial and even essential to good health. Ergo, the ITA proposes that tanning bed use at least during the winter, and perhaps all year for those unable or uninterested in spending ample time outdoors, is required for optimal health in the present and future [12]. This argument, even if one accepts the premise, ignores the fact that vitamin D, identical biochemically and biologically to that derived from cutaneous UV exposure, is naturally present in some foods, added to other "fortified" foods, and readily available as inexpensive oral supplements [11].

The remarkable staying power of this concern that year-round sun exposure may be essential for good health appears to rest on three phenomena: (1) recent evidence that vitamin D may normally play a role in many cells and tissues, beyond its classic role in calcium homeostasis and musculoskeletal health [10]; (2) a strong sentiment among many that a "natural" source of vitamin D is best; and (3) the very complex and highly politicized epidemiologic literature showing in some cases statistical associations between low vitamin D levels and incidence or prevalence of certain diseases in the population [11].

8.2.1 *Vitamin D Nomenclature and Biology*

Exposure of skin to ultraviolet B (UVB) radiation (290–315 nm) converts 7-dehydrocholesterol, a cholesterol precursor, in cell membranes to pre-vitamin D_3. Pre-vitamin D_3's atomic arrangement is subsequently thermally altered to vitamin D_3 (also known as cholecalciferol). Cholecalciferol then binds to a carrier protein and enters the bloodstream. Cholecalciferol is subsequently hydroxylated (a hydroxyl group (–OH) is introduced into the compound) in the liver and kidney to the biologically active form: 1, 25-dihydroxy- vitamin D (1, 25 $(OH)_2$ D) [10]. This active form is tightly regulated and its levels are very constant. 1,25$(OH)_2$D has a cognate nuclear receptor, a class of proteins that regulates gene expression, and functions as a hormone.

Vitamin D also has a far more abundant inactive storage form, 25-hydroxyvitamin D (25(OH)D), the product of a single hydroxylation in the liver. 25(OH)D is easily measured in serum and understood to reflect total body stores of vitamin D. It is now also known that several tissues other than the kidney express the enzyme 1-hydroxylase necessary to convert the storage form of vitamin D to the active form. This has led to the unproven concept that high circulating 25(OH)D levels, in addition to the long-recognized regulatory mechanisms that govern serum 1,25 $(OH)_2$D levels, may influence levels of the active vitamin D within tissues by "mass action." Mass action implies that high levels of a substance in the blood equilibrate with higher intracellular tissue levels and thus, in this case, expose the storage form of the vitamin to the last step in its intracellular activation.

8.2.2 *Vitamin D Levels*

Table 8.1 describes the three classical categories for vitamin D levels [10, 13]. Deficient vitamin D levels cause bone disease in children (rickets) and adults (osteomalacia). The biologic vitamin D deficiency level is strongly influenced by both dietary calcium ingestion and absorption because the functions of vitamin D that determine related signs and symptoms require calcium [14]. Patients with vitamin D deficiency also have an elevated level of parathyroid hormone (PTH), which, like vitamin D, plays a role in the body's calcium homeostasis by stimulating calcium absorption in the intestines and kidneys. Conversely, patients with toxic levels of vitamin D have associated hypercalcemia, hypercalciuria or other disorders associated with high levels of calcium [10].

Table 8.1 Categories of vitamin D levels

1. Deficient	<10–20 ng/mL or 25–50 nmol/L
2. Normal	Any level between deficient and toxic
3. Toxic	>150 ng/mL or 375 nmol/L

Two new and less clearly defined terms have been introduced to describe vitamin D levels: *Insufficient* is defined as a level above "deficient" (described above) but below "sufficient." *Sufficient* levels have been determined by dividing study populations into quartiles or quintiles based on 25(OH)D levels and comparing the incidence or prevalence of the disease of interest in the top and bottom groups. These cut-off values vary among studies. The studied diseases include various kinds of cancer, diabetes, cardiovascular disease, multiple sclerosis, and many other disorders. In some of these studies the selected disease was found to be increased 1.5–2-fold among patients in the lowest group relative to the highest group [15, 16]. Of note, the percentage of affected individuals is quite low regardless of vitamin D level, typically in the range of 0.5–1% in the best group and only slightly more likely in the worst group [17, 18].

The overwhelming majority of "insufficient" subjects do not have the disease of interest. Furthermore, efforts to reduce disease risk among "insufficient" subjects by vitamin D supplementation even over a period of years have been almost entirely unsuccessful, aside from modestly reducing falls and bone fractures in some studies of frail elderly, some of whom were initially vitamin D deficient, rather than insufficient, by conventional criteria [14]. The associational nature of the identified risks therefore raises the concern that the findings are confounded by obesity, known to lower the serum level of 25(OH)D, that, as a fat soluble molecule, dissolves in body fat [11]. Another confounder may be an inactive indoor lifestyle or other factors that may predispose to certain diseases and only incidentally affect 25(OH)D levels [11]. These limitations of the data were emphasized in the recent much-awaited Institute of Medicine (IOM) Report and led that expert panel to caution against large supplements, noting that a benefit of 25(OH)D levels above 20–30 ng/mL (50–75 nmol/L) were not supported by available data and may involve as yet unidentified health risks [19]. Nevertheless, a 2006 consensus conference of five investigators recommended 30–40 ng/mL (75–100 nmol/L) be used as a cutoff for sufficient levels of vitamin D in the population generally [20]. Far higher cutoffs have also been proposed [11].

By these definitions, upwards of three-fourths of many populations of apparently healthy people may be classified as vitamin D insufficient [10, 21], a term now often used interchangeably with "deficient." For example, a frequently visited tanning related website writes, "New research has shown that vitamin D deficiency is epidemic in American adults today, suggesting that up to 90% of North Americans are vitamin D deficient and that vitamin D deficiency has significant implications on human health" [12]. Concern for postulated health risks among those with insufficient vitamin D levels has led many authorities to propose that essentially everyone take vitamin D supplements of at least 1000 international units (IU) daily [11]. This is well above the current governmental recommendation for adequate intake (AI) for vitamin D of 200–600 IU/day depending on one's age [22], recently revised upward to 400–800 IU/day by the IOM panel [19] guidelines developed based on the classic definition of "normal" vitamin D levels. Of note, most multivitamins contain 400 IU of vitamin D. To combat this debatable epidemic of vitamin D deficiency, the indoor tanning industry recommends increased UV exposure (indoor tanning and unprotected sun exposure) as the best way to correct vitamin D insufficiency despite the associated risks of such exposures, that include melanoma, other skin cancers and irreversible photoaging manifest by dryness, wrinkling and irregular pigmentation [8, 23–26]. The ITA argues, "because research suggests that the risks associated with sun exposure are most likely related to intermittent sunburns, it is credible to believe that the benefits of regular, moderate non-burning exposure outweigh the easily manageable risks associated with overexposure… What's more, research has shown that people who utilize indoor tanning equipment that emits UVB – which most tanning equipment does – also produce vitamin D. And studies have also shown that indoor tanning patrons have higher vitamin D blood levels than non-tanners" [12]. It is easy to imagine misinterpretation of these carefully crafted words, particularly by the target population of the tanning industry: young adults and adolescents. This population of young indoor tanners is typically composed of people who tan poorly, are at the lowest statistical risk of vitamin D deficiency, are most likely to achieve generous vitamin D production during incidents of protected sun exposure, and have the highest risk of long-term damage from indoor tanning-induced UV radiation [9].

Those population groups at highest risk of vitamin D deficiency (or insufficiency) rarely frequent indoor tanning facilities: dark-skinned, elderly and/or obese persons. This fact alone would reduce the public health value of indoor tanning as a means of increasing vitamin D "sufficiency" in the U.S., were such a benefit documented. However, in addition, each of these groups experiences a lesser increase in 25(OH)D levels following UVB exposure than lean fair-skinned young adults [27–30]. In dark-skinned individuals, melanin absorbs UVB radiation before it can cause vitamin D photosynthesis; elderly skin also photosynthesizes less vitamin D, likely because epidermal atrophy decreases the amount of vitamin D precursor available; and in obese and overweight individuals, fat soluble vitamin D is sequestered in body fat, removing it from the circulation.

Further, many indoor tanning devices emit primarily UVA. Newer "high pressure" tanning bulbs emit as much as 99–100% UVA radiation in order to increase the speed and duration of tans while decreasing the likelihood of some negative effects such as skin burning and dryness. However, UVA does not lead to vitamin D production. Hence, many new high pressure tanning bulbs will produce no vitamin D whatsoever [5].

Interestingly, few data document a relationship between generous UV exposure and sufficient vitamin D levels. In one study Binkley and colleagues (2007) evaluated a convenience sample of 93 healthy young males from Hawaii in order to assess the relationship [31]. The subjects had a self-reported *minimum* outdoor sun exposure of 15 h per week (mean of 29 h per week) during the 3 months preceding the study. Almost half reported never using sunscreen and the group as a whole reported an average of more than 22 h of unprotected sun exposure per week. Interestingly, the mean 25(OH)D level of the group was 79 nmol/L and more than half had a level under the customary 75 nmol/L "sufficient" cutoff [31].

It has also been well-documented that vitamin D and calcium work together to promote bone health [10], and serum calcium levels are completely unaffected by UV exposure. This again highlights the advantage of a multivitamin containing calcium and vitamin D over indoor tanning for those concerned about the potential health benefits these supplements may provide.

8.2.3 The Bottom Line on Vitamin D Supplementation

In short, no one truly knows the importance of maintaining vitamin D levels above what is classically deficient. Until randomized prospective placebo-controlled studies are performed, no one will know. High doses of oral vitamin D appear to be safe, as has been suggested [10], but they are not without potential adverse effects. In fact, some renal and endocrine authorities warn that there may be increased health risks such as kidney stones associated with large vitamin D supplements, as has already been observed even in populations ingesting conventional supplements of 400 IU/day [32].

Regardless of the true health effects of vitamin D, no data support a claim that sun beds are a better source than oral supplementation. It is known that tanning bed users place themselves at increased risk for skin cancer and premature aging [25]. Finally, tanning beds, with their inherent risks to human health, typically cost $3–15 for a single session and oral vitamin D supplements of 1000 IU daily are as little as $1.50 per month [33]. When one puts the entire picture together, if increasing the level of vitamin D is truly important, the safer and more cost effective approach is obvious.

8.3 Myth #2: Indoor Tanning Is Safer Than Suntanning

The concept that indoor tanning is safer than suntanning arose from the mistaken belief that UVA (320–400 nm) wavelengths did not damage DNA but could stimulate melanogenesis (the production of melanin) in human skin. This led to the creation of tanning lamps that emitted almost entirely UVA and were advertised as giving a safe tan. However, subsequent research established that UVA does cause DNA damage, both directly by the chemically linking adjacent pyrimidine bases and indirectly by creating reactive oxygen species, molecules that then damage guanine bases on the DNA strand [34, 35]. Indeed, tanning is a DNA damage response, and UV irradiation leads to sunburn and tanning in almost exact proportion to its damaging effect on DNA [36–39]. UVB photons are relatively efficient in both regards; depending on their exact wavelengths, UVA photons may be less efficient by a factor of 10,000 or more. In order to induce tanning in response to UVA, tanning lamps were engineered to emit UVA light that is up to 12-fold more intense than natural sunlight [40]. Such lamps may indeed tan (and burn) skin, but do not produce vitamin D, as the action spectrum for vitamin D photosynthesis does not extend into the UVA range and requires UVB energy [10].

The action spectrum for photoaging in human skin, a chronic rather than acute response, is far less well known than the spectra for DNA damage, tanning, burning and vitamin D production. However, based on studies employing mouse models and surrogate end points in human skin, photoaging is widely believed to have substantial contributions from both UVB and UVA [41–43]. Thus, premature skin aging is a further expected adverse effect of tanning bed use, regardless of the UV spectrum of the bulbs employed.

In summary, tanning in a genetically capable individual due to either UVB or UVA radiation reflects DNA damage, and the source of these photons (sunlight or a tanning bed) does not affect this reality. However, the greater intensity (mW/cm^2) of UV in tanning bed lamps compared to sunlight makes it easier to "overdose" and sunburn if an exposure is extended even very briefly beyond the intended "safe" exposure length. UVB intensity in such lamps is frequently 3 times that in equatorial noon sunlight and UVA intensity is up to 12 times higher [44]. On a practical level, more than half of tanning bed users report sunburns [45].

8.4 Myth #3: A Base Tan Before Vacation Reduces Overall Sun Damage

Many of the greater than one million indoor tanners each day are seeking a pre-vacation or "base" tan [25]. This refers to the use of tanning beds to "prepare" skin for a sunny vacation [6]. The use of indoor tanning beds for this purpose is promoted by the ITA, that states, "A tan is the body's natural protection against sunburn. Your skin is designed to tan as a natural body function. Each year, millions of Americans visit professional indoor tanning facilities in the spring, prior to sun-filled vacations or outdoor summertime activities, to establish what tanners know as a "base tan." Doing so enables vacationers to gradually increase their exposure to ultraviolet light without burning" [46].

For those genetically capable of tanning, this ITA strategy may indeed reduce the chance of a sunburn during the first days of intense outdoor sun exposure. Of equal or greater perceived benefit to many patrons, this practice avoids the need to appear in public untanned in a bathing suit or other revealing attire. As noted above, however, the obligate DNA damage is still incurred in achieving the base tan. Moreover, to the extent the "base tan" allows more total UV exposure and reduces subsequent sunscreen use over the vacation period, it may result in greater long-term photodamage. Contrary to popular belief, there is no evidence that pre-vacation tans via tanning bed exposures reduce photoaging and photocarcinogenesis, and the opposite is likely to be the case [6].

Epidemiologic studies have suggested a link between the development of melanoma and sunburns at a young age, and many commercial sunlamp users in Europe and North America report sunburns [23, 47]. Women who use tanning beds more than once a month appear to be significantly more likely to develop melanoma than are less frequent users; and a recent meta-analysis led the International Agency for Research on Cancer (IARC) Working Group on artificial ultraviolet (UV) light and skin cancer to conclude that tanning beds are a human carcinogen [48, 49]. Given the recent increase in skin cancers among young people, particularly young women, it is notable that 70% of indoor tanners are women and that most are aged 16–29 [50–52]. In combination, these findings call into question the "protective effect" that some claim indoor tanning are provides. Indeed, the rather compelling evidence to the contrary has led the FDA to consider reclassifying tanning beds from Class I to Class II or III devices that would require greater controls to ensure safety, to ban their use in people of Fitzpatrick Skin Type I (very fair), and to restrict their use by adolescents [53].

8.5 Myth #4: Indoor Tanning Is a Safe, Effective, and Less Expensive Alternative to Physician Supervised Phototherapy

Numerous cutaneous diseases such as psoriasis and eczema are safely and effectively treated with office-based phototherapy, but the inconvenience of multiple frequent office visits, usually during the work day, is a major obstacle for many patients [54]. Additionally, with the rising costs of healthcare and the associated increase in patient co-payments, many patients are reluctant to use physician-directed light-based therapy [54]. As an alternative, some patients resort to indoor tanning for treatment for their skin condition, in some cases without even an initial physician visit. A U.S. survey of 113 psoriasis patients who had used commercial tanning beds reported that 68% believed it to be an effective treatment modality, although few data exist regarding indoor tanning salons as a means to treat skin disease [55].

One study consisting of both a retrospective chart review of 26 psoriasis patients and a prospective study of 17 patients examined the safety and efficacy of tanning beds in conjunction with oral retinoid medication [56]. In the prospective arm, factors such as tanning salon, tanning bed type, and amount of UVB and UVA dosage were monitored and controlled. In the retrospective study, these factors were largely unknown. Both the retrospective chart review and the prospective trial observed similar safety and efficacy, with 59% of patients in one arm and 82% in the other arm, respectively, clear or almost clear of psoriasis by the end of the study [56]. Nevertheless, the authors suggest that tanning beds should be reserved for patients who do not have the option of prescription phototherapy because of the inherent risks in sending patients to facilities with untrained personnel and unknown UV dosage, intensity, and consistency. As well, the concomitant use of acitretin (the retinoid medication) with the physician-prescribed tanning bed exposure undoubtedly contributed to the investigators' good results.

The variable and usually unknown spectral output from tanning beds also complicates their use as therapeutic devices. Whether the bulbs emit considerable UVB or almost exclusively UVA, however, they can be guaranteed not to emit so-called narrow band (nb) UVB, with peak output at approximately 313 nm [5]. These are the wavelengths shown to be maximally effective for clearing psoriasis without burning and found to be superior to broad-band UVB for treating many other UVB-responsive skin diseases [57–60]. Phototherapy cabinets emitting nb UVB are widely, although not universally, available in dermatology offices and hospital facilities. Using indoor tanning salons as an alternative to physician-supervised nb UVB phototherapy will not only expose the user to the many risks outlined above, but may also lead to a suboptimal treatment response and relatively more exposure to injurious radiation. Hence, available information supports the belief that indoor tanning bed use can improve UV-responsive skin diseases and is often less expensive for users on a per-session basis than physician-monitored office-based phototherapy. However, logic dictates that it is less safe and less effective, a large non-monetary price to pay.

8.6 Summary

Indoor tanning has been controversial for decades and will undoubtedly continue to be so for years to come, pitting immediate cosmetic considerations against future skin health. The ITA has a powerful lobbying presence and is resourceful in articulating the most appealing and most profitable possible case for indoor tanning. The much over-stated individual and public health benefits of UVB-induced vitamin D production is likely to remain central in the industry's fight against government regulation. Niche indications such as establishing a "base tan" and treating certain skin diseases will surely also be cited as appropriate pretexts for cosmetic tanning. To the extent certain groups prize a tanned appearance above the short-term and long-term health of their skin, debunking the myths may have limited impact. One can only hope, however, that some potential indoor tanners will review the facts and make informed decisions before partaking in this misguided behavior.

References

1. Segrave K (2005) Suntanning in 20th century America. McFarland & Company, Inc., Jefferson
2. Albert MR, Ostheimer KG (2002) The evolution of current medical and popular attitudes toward ultraviolet light exposure: part 1. J Am Acad Dermatol 47(6):930–937
3. Martin JM, Ghaferi JM, Cummins DL et al (2009) Changes in skin tanning attitudes. Fashion articles and advertisements in the early 20th century. Am J Public Health 99(12):2140–2146
4. Randle HW (1997) Suntanning: differences in perceptions throughout history. Mayo Clin Proc 72(5):461–466

5. Woo DK, Eide MJ (2010) Tanning beds, skin cancer, and vitamin D: an examination of the scientific evidence and public health implications. Dermatol Ther 23(1):61–71
6. Levine JA, Sorace M, Spencer J et al (2005) The indoor UV tanning industry: a review of skin cancer risk, health benefit claims, and regulation. J Am Acad Dermatol 53(6):1038–1044
7. Indoor Tanning Association (2009a) http://www.theita.com/. Accessed 28 Nov 2009
8. Indoor Tanning Association (2009b) Positive effects of UV light. http://www.theita.com/?page=Positive_Effects_UV&hhSearchTerms=positive±and±effects±and±of±and±UV. Accessed 2 Jan 2010
9. Gilchrest BA (2008) Sun exposure and vitamin D sufficiency. Am J Clin Nutr 88(2): 570S–577S
10. Holick MF, Vitamin D (2007) deficiency. N Engl J Med 357(3):266–281
11. Reddy KK, Gilchrest BA (2010) What is all this commotion about vitamin D? J Invest Dermatol 130(2):321–326
12. Tanning Truth (2010) Vitamin D: one big reason we need sun exposure. http://www.tanningtruth.com/index.php/vitamin_d_sunshine_vitamin. Accessed 3 Jan 2010
13. Schoenmakers I, Goldberg GR, Prentice A (2008) Abundant sunshine and vitamin D deficiency. Br J Nutr 99(6):1171–1173
14. Wolpowitz D, Gilchrest BA (2006) The vitamin D questions: how much do you need and how should you get it? J Am Acad Dermatol 54(2):301–317
15. Giovannucci E (2005) The epidemiology of vitamin D and cancer incidence and mortality: a review (United States). Cancer Causes Control 16(2):83–95
16. Melamed ML, Michos ED, Post W et al (2008) 25-hydroxyvitamin D levels and the risk of mortality in the general population. Arch Intern Med 168(15):1629–1637
17. Freedman DM, Chang SC, Falk RT et al (2008) Serum levels of vitamin D metabolites and breast cancer risk in the prostate, lung, colorectal, and ovarian cancer screening trial. Cancer Epidemiol Biomarkers Prev 17(4):889–894
18. Wactawski-Wende J, Kotchen JM, Anderson GL et al (2006) Calcium plus vitamin D supplementation and the risk of colorectal cancer. N Engl J Med 354(7):684–696
19. Ross AC, Manson JE, Abrams SA et al (2010) The 2011 report on dietary reference intakes for calcium and vitamin D from the Institute of Medicine: what clinicians need to know. J Clin Endocrinol Metab 96(1):53–58
20. Bischoff-Ferrari HA, Giovannucci E, Willett WC et al (2006) Estimation of optimal serum concentrations of 25-hydroxyvitamin D for multiple health outcomes. Am J Clin Nutr 84(1):18–28
21. Ginde AA, Liu MC, Camargo CA Jr (2009) Demographic differences and trends of vitamin D insufficiency in the US population, 1988–2004. Arch Intern Med 169(6):626–632
22. American Academy of Dermatology and AAD Association (2009) Position statement on vitamin D. http://www.aad.org/Forms/Policies/Uploads/PS/PS-Vitamin%20D.pdf. Accessed 6 Jan 2010
23. National Cancer Institute (2008) Melanoma cancer risk. http://www.cancer.gov/melanomarisktool/melanoma-cancer-risk.html. Accessed Mar 2010
24. US Food and Drug Administration (2008) Indoor tanning: the risks of ultraviolet rays. http://www.fda.gov/downloads/ForConsumers/ConsumerUpdates/UCM190664.pdf. Accessed Mar 2010
25. Whitmore SE, Morison WL, Potten CS et al (2001) Tanning salon exposure and molecular alterations. J Am Acad Dermatol 44(5):775–780
26. Karagas MR, Stannard VA, Mott LA et al (2002) Use of tanning devices and risk of basal cell and squamous cell skin cancers. J Natl Cancer Inst 94(3):224–226
27. Arunabh S, Pollack S, Yeh J et al (2003) Body fat content and 25-hydroxyvitamin D levels in healthy women. J Clin Endocrinol Metab 88(1):157–161
28. Clemens TI, Adams JS, Henderson SL et al (1982) Increased skin pigment reduces the capacity of skin to synthesize vitamin D3. Lancet 1:74–76
29. Kumar J, Muntner P, Kaskel FJ et al (2009) Prevalence and associations of 25-hydroxyvitamin D deficiency in US children: NHANES 2001–2004. Pediatrics 124(3):e362–e370, Epub ahead of print

30. Wortsman J, Matsuoka LY, Chen TC et al (2000) Decreased bioavailability of vitamin D in obesity. Am J Clin Nutr 72(3):690–693
31. Binkley N, Novotny R, Krueger D et al (2007) Low vitamin D status despite abundant sun exposure. J Clin Endocrinol Metab 92(6):2130–2135
32. Jackson RD, LaCroix AZ, Gass M et al (2006) Calcium plus vitamin D supplementation and the risk of fractures. N Engl J Med 354(7):669–683
33. CVS (2010) CVS High potency vitamin D. http://www.cvs.com/CVSApp/catalog/shop_product_detail.jsp?filterBy=&skuId=451102&productId=451102&navAction=push&navCount=3&no_new_crumb=true. Accessed 5 Apr 2010
34. Mouret S, Baudouin C, Charveron M et al (2006) Cyclobutane pyrimidine dimers are predominant DNA lesions in whole human skin exposed to UVA radiation. Proc Natl Acad Sci USA 103(37):13765–13770
35. Ren X, Li F, Jeffs G et al (2009) Guanine sulphinate is a major stable product of photochemical oxidation of DNA 6-thioguanine by UVA irradiation. Nucleic Acids Res 38(6):1832–1840
36. Eller MS, Ostrom K, Gilchrest BA (1996) DNA damage enhances melanogenesis. Proc Natl Acad Sci USA 93(3):1087–1092
37. Cui R, Widlund HR, Feige E et al (2007) Central role of p53 in the suntan response and pathologic hyperpigmentation. Cell 128(5):853–864
38. Parrish JA, Jaenicke KF, Anderson RR (1982) Erythema and melanogenesis action spectra of normal human skin. Photochem Photobiol 36(2):187–191
39. Freeman SE, Hacham H, Gange RW et al (1989) Wavelength dependence of pyrimidine dimer formation in DNA of human skin irradiated in situ with ultraviolet light. Proc Natl Acad Sci USA 86(14):5605–5609
40. Epstein JH, Wang SQ (2010) Understanding UVA and UVB. http://www.skincancer.org/understanding-uva-and-uvb.html. Accessed Mar 2010
41. Lavker RM, Gerberick GF, Veres D et al (1995) Cumulative effects from repeated exposures to suberythemal doses of UVB and UVA in human skin. J Am Acad Dermatol 32(1):53–62
42. Kligman LH (1989) The ultraviolet-irradiated hairless mouse: a model for photoaging. J Am Acad Dermatol 21(3 Pt 2):623–631
43. Kang S, Fisher GJ, Voorhees JJ (2001) Photoaging: pathogenesis, prevention, and treatment. Clin Geriatr Med 17(4):643–659, v-vi
44. Miller SA, Hamilton SL, Wester UG et al (1998) An analysis of UVA emissions from sunlamps and the potential importance for melanoma. Photochem Photobiol 68(1):63–70
45. Cokkinides V, Weinstock M, Lazovich D et al (2009) Indoor tanning use among adolescents in the US, 1998 to 2004. Cancer 115(1):190–198
46. Indoor Tanning Association (2010) Indoor tanning FAQs. http://www.theita.com/?page=FAQs. Accessed 5 Jan 2010
47. Coelho SG, Hearing VJ (2010) UVA tanning is involved in the increased incidence of skin cancers in fair-skinned young women. Pigment Cell Melanoma Res 23(1):57–63
48. Horner MJ, Ries LAG, Krapcho M, et al (2010) SEER Cancer Statistics Review, 1975–2006. http://seer.cancer.gov/csr/1975_2006/index.html. Accessed Mar 2010
49. International Agency for Research on Cancer Working Group on Artificial Ultraviolet (UV) Light and Skin Cancer (2007) The association of use of sunbeds with cutaneous malignant melanoma and other skin cancers: a systematic review. Int J Cancer 120:1116–1122
50. Christenson LJ, Borrowman TA, Vachon CM et al (2005) Incidence of basal cell and squamous cell carcinomas in a population younger than 40 years. JAMA 294(6):681–690
51. Purdue MP, Freeman LE, Anderson WF et al (2008) Recent trends in incidence of cutaneous melanoma among US Caucasian young adults. J Invest Dermatol 128(12):2905–2908
52. Swerdlow AJ, Weinstock MA (1998) Do tanning lamps cause melanoma? An epidemiologic assessment. J Am Acad Dermatol 38(1):89–98
53. WebMD Health News (2010) FDA Panel:New tanning bed restrictions needed. http://www.webmd.com/skin-beauty/news/20100330/fda-panel-new-tanning-bed-restrictions-needed. Accessed 5 Apr 2010

54. Yentzer BA, Feldman SR (2009) Tanning beds as an alternative for psoriasis when office-based phototherapy is not accessible. J Dermatolog Treat 20(4):238–240
55. Fleischer AB Jr, Feldman SR, Rapp SR et al (1996) Alternative therapies commonly used within a population of patients with psoriasis. Cutis 58(3):216–220
56. Carlin CS, Callis KP, Krueger GG (2003) Efficacy of acitretin and commercial tanning bed therapy for psoriasis. Arch Dermatol 139(4):436–442
57. Parrish JA, Jaenicke KF (1981) Action spectrum for phototherapy of psoriasis. J Invest Dermatol 76(5):359–362
58. Habib F, Stoebner PE, Picot E et al (2005) Narrow band UVB phototherapy in the treatment of widespread lichen planus. Ann Dermatol Venereol 132(1):17–20
59. Walters IB, Burack LH, Coven TR et al (1999) Suberythemogenic narrow-band UVB is markedly more effective than conventional UVB in treatment of psoriasis vulgaris. J Am Acad Dermatol 40(6 Pt 1):893–900
60. Degitz K, Messer G, Plewig G et al (1998) Narrow band UVB 311 nm versus broad band UVB. New developments in phototherapy. Hautarzt 49(10):795–806, quiz 806

Chapter 9
A Systematic Review of Intervention Efforts to Reduce Indoor Tanning

Rob Turrisi, Joel Hillhouse, Kimberly Mallett, Jerod Stapleton, and June Robinson

Abstract This chapter reviews the literature examining interventions to reduce indoor tanning (IT). The first objective was to highlight programs that show promise for large scale dissemination. The second objective was to promote criteria and standards for future intervention research efforts. The scope of interest for this review includes universal (for everyone in the population), selective (for those in the population who are at a greater risk), and indicated (for those who already are experiencing conditions that identify them as at risk) programs. The evaluation of the interventions resulted in three levels of evidence: (1) most promising, (2) emerging, and (3) mixed. For an intervention to be considered "most promising", it was required that ten criteria be met through examination of research findings in published reports consistent with Flay and colleagues (Prev Sci 6(3):151–175, 2005). Interventions that were classified as "emerging" met most of the criteria. Finally, interventions classified as "mixed" did not reach threshold on more than two criteria that were deemed critical. The results revealed that there was very limited research on IT interventions that meet all the evaluation criteria. Only one intervention

R. Turrisi (✉) • K. Mallett
Biobehavioral Health & Prevention Research Center, The Pennsylvania State University, 109 S. Henderson, University Park, PA 16801, USA
e-mail: rturrisi@psu.edu; kmallett@psu.edu

J. Hillhouse
Department of Community and Behavioral Health, East Tennessee State University, Box 70674, Johnson City, TN 37614, USA
e-mail: hillhous@etsu.edu

J. Stapleton
The Cancer Institute of New Jersey, UMDNJ-Robert Wood Johnson Medical School, 195 Little Albany Street, New Brunswick, NJ 08901, USA
e-mail: jerod.stapleton@umdnj.edu; jls82@psu.edu

J. Robinson
Department of Dermatology,
Northwestern University Feinberg School of Medicine, Chicago, IL 60609, USA
e-mail: june-robinson@northwestern.edu

C.J. Heckman and S.L. Manne (eds.), *Shedding Light on Indoor Tanning*,
DOI 10.1007/978-94-007-2048-0_9,

approach met all of the criteria (Appearance Booklet) (Hillhouse and Turrisi, Behav Med 25(4):395–409, 2002; Hillhouse et al., Cancer 113(11):3257–3266, 2008). Although the number of published papers in the IT area has increased dramatically over the past decade, these efforts have yet to translate into rigorously conducted intervention trials. The review points to important issues that need to be addressed in future research on the prevention of IT.

Keywords Indoor tanning • Interventions • Evidence • Appearance Booklet • UV photo • Motivational interviewing • Robert Turrisi • Joel Hillhouse • Karen Mallett • June Robinson • Jerod Stapleton • Brian Flay • Frederick Gibbons • Review • Young adults

Abbreviations

IT	Indoor tanning
MI	Motivational interviewing
RCT	Randomized controlled trial
US	United States
UV	Ultraviolet

9.1 Introduction

Much of the published work on indoor tanning (IT) begins with a description of the public health problem posed by skin cancer, epidemiological data on the prevalence of skin cancer, the consequences of skin cancer, and the association between skin cancer and IT. It is apparent from these reports that the problem of skin cancer is considerable and IT rates are highest for adolescent, emerging adult, and young adult females [1]. Recent reviews indicate approximately 10% of United States adolescents under the age of 15 have engaged in IT in the past year, with usage rates increasing to 25–40% in older adolescent females and young adults. Using conservative estimates, individuals who start tanning before age 15 will likely accrue at least 50–100 IT sessions in their lifetime [2, 3]. Furthermore, there is evidence that early initiation of IT is associated with long-term habitual indoor tanning behavioral patterns later in life, which are highly resistant to change. It is also noteworthy that such problematic levels of IT occur in developed countries worldwide [4].

Perhaps the single most important point to be made about IT is that there can be substantial consequences, including mortality from melanoma and nonmelanoma skin cancer, that take a tremendous toll on individuals and families. Age of IT initiation is associated with a risk of developing melanoma after 40 and nonmelanoma skin cancer after 50 [5]. Nonmelanoma skin cancer in Caucasians increases with age, which appears to be due in part to cumulative environmental ultraviolet (UV) radiation [6]. Survival is improved with early detection and surgical treatment [7]. In those diagnosed with Stage IA melanoma, the survival rate 10 years after diagnosis was estimated to be higher than 95%, however this declined to less than 60% when diagnosis occurred later (Stage IIB, C) [7]. People who are diagnosed later rapidly progress to the advanced stage of the disease with metastasis to internal organs (Stage IV). Melanoma, the second most common cancer diagnosed in young women, remains an important cause of mortality in the US with nearly 8,000 US deaths per year [8].

The prevalence rates of skin cancer and its association with intentional UV exposure such as IT warrant a comprehensive public health approach firmly grounded in evidence-based preventive interventions and policy making. From a public health perspective, there are many challenges in addressing IT and skin cancer. Our view is that the major challenges are the systematic design and evaluation of interventions across the critical developmental stages where IT is most prevalent. There is also a greater need for effective individual and community interventions as well as policies that regulate use. The latter is addressed in the subsequent chapter (Chap. 10, Suarez et al.). The former is the focus of this review. Thus, the objectives of this review are twofold. The first objective is to highlight existing programs that show promise for larger scale dissemination. The second objective is to promote criteria and standards for future intervention efforts in the field.

9.2 Methods

9.2.1 Intervention Selection Criteria

The scope of interest for this review includes universal (for everyone in the population), selective (for those in the population who are at a greater risk), and indicated (for those who already are experiencing conditions that identify them as at risk) prevention programs. Two types of literature were reviewed to ensure all relevant evidence on IT interventions was included. First, given the standards in the peer review process, refereed professional journals were used. The search of databases for empirical peer reviewed papers included Medline, PsycINFO, Science Citation Index Expanded, and the Social Science Citation Index. Second, relevant Internet sources were checked including the web pages for the National Institutes of Health, Center for Disease Control and Prevention, the American Medical Association, and the Society for Prevention Research. From these collective sources, eight reports were discovered covering five different types of interventions. Finally, we contacted the originating research teams during the review process to address specific questions for accuracy and to determine if there were additional reports that might have been missed.

9.2.2 Intervention Evaluation Criteria

A set of criteria was used to evaluate the interventions identified in the search process. The evaluation of the interventions resulted in three levels of evidence: (1) most promising, (2) emerging, and (3) mixed. For an intervention to be considered "most promising", it was required that ten criteria be met through the examin of the research findings in published reports consistent with Flay and colleagues' standards of evidence for efficacy, effectiveness, and dissemination, and the National Registry of Evidence Based Programs and Practices [9]. The first criterion concerned the experimental design. Studies were evaluated on whether they used a randomized controlled trial (RCT) or a quasi-experimental design that had an adequate comparison group in the implementation and evaluation of the intervention. The second criterion entailed having two rigorous trials conducted to establish the reliability of the efficacy of the intervention. The third criterion was that the sample in the intervention trial was randomly selected from the target population of interest (or risk). The fourth criterion was that there needed to be random assignment of study participants to the treatment and control conditions. The fifth criterion was that the outcome measures had to assess indoor tanning behaviors and not proxies (e.g., intentions to tan or attitudes about tanning). The sixth criterion required that evidence was provided showing that the outcome measures used were reliable and valid. The seventh criterion was that rigorous statistical approaches were applied to evaluate treatment effects (e.g., statistical and clinical significance between the treatment and control/comparison groups and effect sizes). The eighth criterion required that positive effects on behavioral outcome measures were reported (e.g., differences in IT quantity or frequency between the conditions). The ninth criterion was that there were no iatrogenic (negative effects caused by the intervention such as increased indoor tanning) or unanticipated effects observed or reported. The final criterion was that at least a 6-month post-baseline follow-up assessment was conducted to ensure the treatment effects were not short-lived. Interventions that were classified as "emerging" met most of the criteria, but did not reach threshold on one or two that were deemed critical (e.g., long-term follow-ups). Finally, interventions classified as "mixed" did not reach threshold on more than two criteria that were deemed critical.

9.3 Results

Interventions in the three classes of evidence are summarized in Table 9.1. To be included in the table it was *necessary* that the interventions at least met the criteria of having: (1) an experimental design where either a randomized trial or a quasi-experimental design that used an adequate comparison group, (2) outcome measures assessing indoor tanning behaviors, and (3) positive effects on behavioral outcome measures. Thus, the remaining seven criteria were used to classify interventions as most promising, emerging, or mixed. Only the former two classifications were reviewed in detail.

9.3.1 Most Promising

One intervention met all seven remaining criteria: The Appearance Booklet [10, 11]. The Appearance Booklet is 27 pages in length. Section 9.1 provides an introduction

Table 9.1 Intervention studies and criteria for evaluation

Classification	Type of intervention	Citation	Two rigorous trials	Random selection	Random assignment	Reliable and valid measures	Rigorous statistical approach	Iatrogenic effects	6 month + follow-up
Most promising	Appearance Booklet	Hillhouse and Turrisi [10]	Y	Y	Y	Y	Y	N	N
		Hillhouse et al. [11]	Y	Y	Y	Y	Y	N	Y
		Stapleton et al. [12]	Y	Y	Y	Y	Y	N	Y
Emerging	UV photograph	Gibbons et al. [13]	N	Y	Y	Y	Y	N	N
	Motivational interviewing	Turrisi et al. [14]	N	N	Y	Y	Y	N	N
Mixed	Message framing	Greene and Brinn [15]	N	N	np	Y	Y	N	N
	Mortality salience	Routledge et al. [16]	N	N	Y	N	Y	N	N
		Cox et al. [17]	N	N	Y	Y	Y	N	N

Y Yes, *N* No, *np* not provided in manuscript

on the history of tanning and a discussion of current fashion trends moving toward paler, unblemished, and untouched skin. The next section provides an in-depth discussion of the effects of ultraviolet radiation on the skin, differences between UVA and UVB radiation, and the role of UVA radiation in premature skin aging and wrinkling. The next section encourages readers to give up indoor tanning altogether but follows a harm-reduction strategy emphasizing IT less than 10 times per year, not tanning outside after IT, not trying to maintain a tan year round, not tanning in the nude, always wearing protective goggles when IT, and using sunless tanning products as alternatives to UV exposure. Because the intervention was in part based on decision-making theory that emphasizes attitudes toward IT, alternatives to IT, and individuals' behavior being a choice among the most preferred alternatives available, the last section provided information on a number of appearance-enhancing alternatives to tanning including exercise, fashion, and sunless tanning products.

Two independent rigorous trials have been conducted on the Appearance Booklet, and in both cases reductions in IT were reported with no evidence of iatrogenic effects [10, 11]. Hillhouse and Turrisi randomly recruited 147 female college students from the general student body at a moderate size southeastern university who reported indoor tanning at least monthly into a RCT at the start of the winter semester (January/February) [10]. Two months post-baseline (and following implementation of the intervention), participants received a follow-up assessment of indoor tanning beliefs, attitudes, intentions, and IT behavior in the past 2 months (72% retained at follow-up). The intervention significantly reduced all IT outcome measures relative to no treatment controls with intervention participants reporting approximately half the number of IT visits as controls, and less than half the number of IT visits they reported at pre-intervention despite the fact that follow-up occurred during the heaviest indoor tanning period (i.e., March to April) [effect size (Cohen's d) = 0.35].

Hillhouse and colleagues replicated the earlier results in a larger sample, which included a longer-term follow-up, validated outcome measures, electronic diaries, and a design that allowed for mediation analyses to understand the processes underlying the intervention effects [11]. The investigators recruited 430 young adult female undergraduate indoor tanners at the start of the fall semester who were assessed at baseline, a 3-month post intervention, and again at a 9-month follow-up (participation rates were 95%, and attrition rates across the 9 month study were less than 5%). At all three assessments, respondents completed measures assessing their attitudes toward IT (e.g., I feel favorable about indoor tanning), sunless tanning, and using clothing as an alternative to tanning for appearance enhancement, perceptions that tanning is attractive and relaxing, normative beliefs (e.g., whether female students should IT), perceived susceptibility to skin damage and past 3 month IT use. Observation of the effects indicated that the intervention reduced IT visits relative to no treatment controls by over 35% [effect size (Cohen's d) = 0.29]. Mediation analyses examining the effects of the intervention on the attitudinal items at the short-term follow-up and then the effects of these items on IT at the long-term follow-up revealed that the intervention also significantly altered attitudes toward IT, sunless tanning, and using clothing as an alternative for appearance enhancement, perceived susceptibility to skin damage, and perceptions that tanning is attractive, relaxing, and normative. In turn, these attitudinal and cognitive constructs significantly influenced IT visits, providing evidence that the intervention affected core motivational mediators and IT behavior over a long follow-up period.

9.3.2 Emerging Evidence

The first approach that was classified as emerging evidence was the UV photograph approach reported in Gibbons and colleagues [13]. The UV photograph approach consisted of providing participants with a photo of themselves taken with a camera with a UV filter. The photo shows skin damage that is normally invisible to the naked eye.

Two studies were conducted which revealed reduced IT without evidence of iatrogenic effects. In study one, 80 male and female undergraduate students were randomized to one of two conditions (UV photo, no UV photo) and were compared on IT at a 4-week follow-up. Individuals in the UV photo condition were also provided with information regarding the risks of UV exposure. In study two, 134 male and female undergraduate students were again randomized to one of two conditions (UV photo, no UV photo) and assessed at 4 weeks, but this time both groups were provided with information regarding the risks of UV exposure. Although these studies had random selection, random assignment to conditions, reliable measures of IT constructs (e.g., appearance beliefs, susceptibility; all alphas of these constructs were greater than 0.7), and rigorous statistical analyses assessing statistical significance and effect sizes, the follow-up period was too short to determine whether the effects were sustained or short-lived. Thus, despite the promise of the approach, it was classified under the emerging evidence category.

The other approach classified as emerging evidence was the Motivational Interviewing (MI) approach [14]. Motivational interviewing is a client-centered approach originally developed in the addictive behaviors field with the intent of promoting and supporting behavior change. MI theory argues that solutions for changes in behavior generated by clients (e.g., change talk) in collaboration with their providers have a greater probability of being carried out than solutions generated by providers only. In MI, the provider accepts the client's ambivalence about changing behaviors that have a perceived mixture of benefits and risks, like intentional tanning.

In a study by Turrisi and colleagues, the MI condition consisted of a 30-minute one-on-one peer counseling intervention based primarily upon personalized feedback (provided orally and via feedback sheets) from participants' baseline assessment information on IT attitudes, beliefs, norms, and behaviors [14]. MI sessions were conducted by trained undergraduate peer counselors receiving course credit for their participation. Peer counselors were trained in MI and basic counseling skills. Training involved over 50 hours of MI, counseling, cognitive-behavioral skills and content area (skin cancer) material and was conducted by an advanced doctoral student with over 6 years of related experience. In addition, peer counselors completed multiple role-plays in which practice sessions were audio recorded and reviewed for MI and content adherence. MI and protocol content adherence was assessed by scoring the recorded sessions using the Motivational Interviewing Treatment Integrity Code: Version 2.0. Scores of 5 or higher on a 1–7 scale (with 7 being highest) met the requirements of being adherent. For example, protocol-specific items pertained to information discussed with participants from their feedback sheets related IT to negative short- and long-term risks and consequences, tanning norm comparisons, challenging tanning expectancies, and exploration of healthy alternatives to indoor tanning (e.g., spray on tanning). All participants received a copy of their feedback sheets upon completion of the interview.

One hundred and five female undergraduate students who reported indoor tanning more than 10 times in the previous year were randomized into the MI group ($n=39$), a mailed feedback only group (e.g., the same feedback sheets as given the MI group, but without the MI counselor interview) ($n=34$), or no treatment control group ($n=32$). Participants in the MI group reported significantly fewer indoor tanning sessions ($M=4.41$, SD=7.74) at the 3-month post-baseline follow-up relative to both the control ($M=11.78$, SD=13.03) and the mailed feedback groups ($M=9.03$, SD=11.92) [effect size (group by session interaction Cohen's d)=0.07]. No significant differences were observed between the mailed feedback group and the controls. There were a number of positive methodological aspects and no iatrogenic effects: the inclusion of indoor tanners, random assignment, reliable and valid measures, and rigorous statistical analyses. There were several limitations. First, the study was not replicated, the sample size was small, and the follow-up period was short. Thus, it was classified under the emerging evidence category.

9.3.3 Mixed Evidence

The first approach that was classified as mixed evidence was the Message Framing approach reported in Green and Brinn [15]. The Message Framing approach consists of examining what features of messages are most effective in reducing IT. The two types of messages examined were: (1) statistical information summarizing deaths and health threats from skin cancer, and (2) a narrative of a case study of a particular person with skin cancer. This study revealed reduced IT without evidence of iatrogenic effects [15]. Although the study used reliable and valid measures and rigorous statistical analyses, it utilized a convenience sample, did not appear to use random assignment, and did not have a follow-up period beyond 6 months. Thus, because the science lacked the required methodological rigor and standards of evidence, it was classified under the mixed evidence category.

The second approach that was classified as mixed evidence was the Mortality Salience approach reported in Cox and colleagues [17] and Routledge and colleagues [16]. The Mortality Salience approach consists of reminding individuals of their own mortality by having individuals complete a 15-item survey that increases death-related thoughts. In doing so, individuals theoretically reduce their anxiety by living up to culturally derived standards (e.g., tanned skin is attractive). In the several studies conducted, positive effects were observed; although the studies tended to rely on convenience samples, short follow-up periods and proxies of IT. Thus, because the science lacked the standards of evidence in Flay and colleagues, it was classified under the mixed evidence category [9].

Both of the reports described in this section did not include detail about several of the criteria under review. Future studies should consider implementing and evaluating these interventions with more stringent criteria to see whether these approaches warrant greater attention in the literature.

9.4 Issues in Current and Future IT Intervention Research

Overall, it is noteworthy that there is very limited research on IT interventions that meet the criteria for standards of evidence including two rigorous trials, random selection of participants, and follow-up evaluations of 6 months post-baseline. Only one intervention approach met all of the criteria (the Appearance Booklet) [10, 11]. Although the number of published papers in the IT area has increased dramatically over the past decade, there have been few rigorously conducted intervention trials. The discussion of the reasons why this is the case are beyond the scope of this chapter; however, the current review points to important serious gaps, omissions, and other issues that need to be addressed in future research on the prevention of IT. These will be briefly discussed in turn.

9.4.1 Limited Intervention Approaches

Although there are interventions designed to reduce UV exposure or increase sun-protection, there are few interventions that specifically target IT. This is problematic given the growth of the industry, the usage by adolescent and young adults, and IT's definitive association with skin cancer. This review should serve as a call to both skin cancer researchers and more general prevention researchers to develop new approaches such as community, family, school-based, and health-care delivery interventions to increase the uptake and delivery of prevention efforts in this important health area. There is also a need to utilize multiple modalities of interventions. Our review revealed the benefits of having individuals read a short booklet, look at a photo, or speak to a trained peer. It is important to note that the successful approaches did not involve the use of technologies such as the web and handheld computers. The most successful approaches had theoretical underpinnings as to why individuals engaged in the behavior and what would motivate them to change.

Given the limited number of evidence-based efficacious IT interventions, a more broad-based view in the field is called for to change this risky behavior.

9.4.2 Methodological Rigor

The standards of evidence that we used to evaluate the interventions are the cornerstones of empirical prevention science. Concepts such as replication, random selection, reliable and valid measurement, rigorous statistical evaluation, and long-term assessments are *the* standards of intervention research. Despite this, the studies we reviewed were uneven in their approach and presentation, by and large, consistent with what Dodd and Forshaw found in their review of appearance-based UV reduction interventions using less stringent evaluative criteria than what was used in the present chapter [18]. The good news is that the IT field is young, which will allow for future work to meet the needs of improving the methodological rigor.

9.4.3 Limited Replications

As indicated, there is a tremendous need for replications of the intervention outcome studies. Independent replications by different investigators are very rare in prevention science and are almost nonexistent in the IT area. Further, follow-up studies examining larger or more diverse samples, high versus low risk groups, high versus low risk environments (e.g., where there are lower densities of places where teens can indoor tan), varying dosages of interventions, treatment boosters, different implementation procedures, and longer-follow-up periods are all appropriate methods of assessing replications of findings of earlier studies. To date, there is very little replication operating in the IT domain.

9.4.4 Limited Longitudinal Examinations

There is a need for studies that utilize longitudinal designs with evaluations that extend to 6 months and beyond. It cannot be overstated how important it is to determine whether the effects of the interventions are sustained or short-lived. The typical design in the IT area and reviewed here assessed behavior at 30 days or less. Based on the Standards of Evidence such follow-up periods are simply too short to adequately assess decay of an intervention effect [9]. Further, even among those studies that met the criterion of having greater than or equal to 6 months of follow-up data, there were none that examined the behavior one or more years post intervention. Given that studies of IT patterns have revealed at least some groups that are influenced by seasonal changes, 1 year or longer follow-up assessments seem warranted [19]. Finally, the use of longitudinal designs with multiple assessments permits the expansion of research questions that can be assessed using growth mixture analytic approaches. For example, do different types of tanners (e.g., seasonal vs. habitual) show the same patterns over time in response to various diverse approaches (e.g., universal vs. tailored interventions)? The IT field is young, but such designs are being utilized in other prevention areas to good effect and could be of great utility in reducing IT and the consequences that ensue from this risky behavior.

9.4.5 Other Considerations

Because the field is still in its early phases of development, there remain many content-based questions that are unanswered. For example, there have been no studies conducted that have specifically examined the optimal dose necessary to achieve short-term or sustained behavioral change. Thus, it is unclear how long or short the Appearance Handbook needs to be to achieve a desired amount of behavioral change in IT or whether boosters could improve the duration of the effects. Moreover, although several of the studies demonstrated the benefits of using appearance-focused approaches (Appearance Handbook and UV photographs), the studies focused on different antecedents of behavior change. Future research in this area might benefit from identifying common mediation constructs underlying these diverse intervention efforts. Finally, there have been no systematic efforts that focus on individual differences that serve as moderators of intervention efficacy. It is not clear from the research in the area what are the person-based characteristics (e.g., self-regulation, impulsivity, self-determination) that influence whether interventions work better or worse.

9.5 Conclusions

In sum, the current chapter provided an overview of intervention approaches focused on reducing IT. Even though this is a relatively new area of research, 50% of the intervention work in the field to date was classified as either most promising or emerging. Of those interventions reviewed, ones that focused on appearance or peer counseling using personalized feedback seemed to hold the most promise. While the research to date has formed a foundation in this important area, more work targeting IT is needed. The chapter highlighted ways in which intervention approaches can be developed and evaluated in a more rigorous manner in order to move the field forward, and more importantly, turn promising findings into actual practice.

Acknowledgements The authors would like to thank the American Cancer Society and the National Cancer Institute for their support of the research of Drs. Turrisi, Hillhouse, Mallett, Stapleton, and Robinson, and also Kelly Guttman for her comments on earlier versions of the chapter.

References

1. Geller AC, Colditz G, Oliveria S et al (2002) Use of sunscreen, sunburning rates, and tanning bed use among more than 10 000 US children and adolescents. Pediatrics 109(6):1009–1014
2. Autier P (2004) Perspectives in melanoma prevention: the case of sunbeds. Eur J Cancer 40(16):2367–2376
3. Demko CA, Borawski EA, Debanne SM et al (2003) Use of indoor tanning facilities by white adolescents in the United States. Arch Pediatr Adolesc Med 157(9):854–860
4. Schneider S, Kramer H (2010) Who uses sunbeds? A systematic literature review of risk groups in developed countries. J Eur Acad Dermatol Venereol 24(6):639–648
5. Diffey BL, Langtry JA (2005) Skin cancer incidence and the ageing population. Br J Dermatol 153(3):679–680
6. Titus-Ernstoff L, Perry AE, Spencer SK et al (2006) Multiple primary melanoma: two-year results from a population-based study. Arch Dermatol 142(4):433–438
7. Jemal A, Siegel R, Ward E et al (2008) Cancer statistics, 2008. CA Cancer J Clin 58(2):71–96
8. American Cancer Society (2009) Cancer facts & figures 2008. American Cancer Society, Inc., Atlanta
9. Flay BR, Biglan A, Boruch RF et al (2005) Standards of evidence: criteria for efficacy, effectiveness and dissemination. Prev Sci 6(3):151–175
10. Hillhouse JJ, Turrisi R (2002) Examination of the efficacy of an appearance-focused intervention to reduce UV exposure. J Behav Med 25(4):395–409
11. Hillhouse J, Turrisi R, Stapleton J et al (2008) A randomized controlled trial of an appearance-focused intervention to prevent skin cancer. Cancer 113(11):3257–3266
12. Stapleton J, Turrisi R, Hillhouse J et al (2010) A comparison of the efficacy of an appearance-focused skin cancer intervention within indoor tanner subgroups identified by latent profile analysis. J Behav Med 33(3):181–190
13. Gibbons FX, Gerrard M, Lane DJ et al (2005) Using UV photography to reduce use of tanning booths: a test of cognitive mediation. Health Psychol 24(4):358–363
14. Turrisi R, Mastroleo NR, Stapleton J et al (2008) A comparison of 2 brief intervention approaches to reduce indoor tanning behavior in young women who indoor tan very frequently. Arch Dermatol 144(11):1521–1524
15. Greene K, Brinn LS (2003) Messages influencing college women's tanning bed use: statistical versus narrative evidence format and a self-assessment to increase perceived susceptibility. J Health Commun 8(5):443–461
16. Routledge C, Arndt J, Goldenberg JL (2004) A time to tan: proximal and distal effects of mortality salience on sun exposure intentions. Pers Soc Psychol Bull 30(10):1347–1358
17. Cox CR, Cooper DP, Vess M et al (2009) Bronze is beautiful but pale can be pretty: the effects of appearance standards and mortality salience on sun-tanning outcomes. Health Psychol 28(6):746–752
18. Dodd LJ, Forshaw MJ (2010) Assessing the efficacy of appearance-focused interventions to prevent skin cancer: a systemic review of the literature. Health Psychol Rev 4(2):93–111
19. Hillhouse J, Turrisi R, Shields AL (2007) Patterns of indoor tanning use: implications for clinical interventions. Arch Dermatol 143(12):1530–1535

Chapter 10
Indoor Tanning Regulation, Enforcement, Taxation, and Policy

Andrea Suárez, Robert P. Dellavalle, and June K. Robinson

Abstract Tanning device use before the age of 35 is associated with a 75% increase in the risk for melanoma, a statistic prompting increased efforts towards limiting youth access to indoor tanning facilities. Numerous state, federal, and international restrictions focus on age restrictions, parental consent requirements, ultraviolet (UV) radiation exposure amount and frequency, warning labeling on the devices, taxation, and operator education. While commendable, these efforts are limited by non-compliance and insufficient enforcement of existing regulations.

A parallel can be drawn between the use of tobacco products and indoor-tanning. Despite serious health risks, large industries target teens and young adults to initiate indoor tanning and smoking through advertising that portrays these behaviors as socially appealing. Age restriction policies coupled with taxation effectively limit youth access to tobacco products, and therefore, may be useful models for effective tanning legislation strategies.

Keywords Legislation • Regulation • Taxation • Restrict youth access • Advertising • Food and Drug Administration • Federal Trade Commission • Ultraviolet radiation • Limiting youth access to tobacco products • Public health • Education • Skin cancer • US federal regulations • Indoor Tanning Association • World Health Organization • Tanning beds • Minimal erythema dose

A. Suárez
Department of Dermatology, University of Colorado Denver, Aurora, CO, USA

R.P. Dellavalle
Department of Dermatology, University of Colorado Denver, Aurora, CO, USA

Denver Veterans Affairs Medical Center, Denver, CO, USA

Epidemiology, Colorado School of Public Health, Aurora, CO, USA

J.K. Robinson (✉)
Department of Dermatology, Northwestern University Feinberg School of Medicine,
132 E. Delaware Place, #5806, Chicago, IL 60611, USA
e-mail: june-robinson@northwestern.edu

C.J. Heckman and S.L. Manne (eds.), *Shedding Light on Indoor Tanning*,
DOI 10.1007/978-94-007-2048-0_10, © Springer Science+Business Media B.V. 2012

Abbreviations

AAD	American Academy of Dermatology
AADA	American Academy of Dermatology Association
AMA	American Medical Association
ASDS	American Society for Dermatologic Surgery
ASP	American Society for Photobiology
CDC	Centers for Disease Control and Prevention
FDA	Food and Drug Administration
FTC	Federal Trade Commission
ITA	Indoor Tanning Association
MED	Minimal erythemal dose
SCC	Squamous cell carcinoma
SID	Society for Investigative Dermatology
UK	United Kingdom
UV	Ultraviolet
WHO	World Health Organization

10.1 Introduction

Indoor tanning poses well-established dangers to users, namely an increased risk of skin cancer, as well as photosensitive and phototoxic reactions, ocular damage, and premature wrinkling of the skin [1]. The Food and Drug Administration (FDA) [2], Federal Trade Commission (FTC) [3], the American Academy of Dermatology [4], the Centers for Disease Control and Prevention (CDC) [5], the Department of Health and Human Services [6], the World Health Organization (WHO) [7], and the American Medical Association (AMA) [8] uniformly acknowledge the harms of increased UV radiation exposure, as occurs through indoor tanning, and its causative role in skin cancer.

Despite the mantra that "there is no such thing as a safe tan," several indoor tanning advertisements maintain that indoor tanning does not cause cancer, boasting claims of its safety, and go as far as to promise health benefits, especially with reference to the need for ultraviolet light exposure to maintain adequate levels of vitamin D [9–13]. Inadequate federal and state regulation is one factor responsible for the perpetuation of these advertisements and the unnecessary risk to consumers, particularly adolescents and young adults, as a result of this misinformation. Currently, only 28 states have passed legislation for indoor tanning regulation [14, 15]. Recognizing the need for improvement in tanning salon regulation, in 2006 the AMA resolved to "develop model state legislation to prohibit the sale of tanning parlor ultraviolet rays to those under 18 years of age except as prescribed by a physician and that this model legislation be widely disseminated to the Federation," and "that the Food and Drug Administration's Center for Devices and Radiological Health immediately hold a fair hearing on the safety and efficacy of ultraviolet bulbs as used in indoor tanning facilities and make its findings publicly available [11]."

10.2 Restrictions on Indoor UV Tanning Access

Melanoma is the second and third most common cancer among young adult women and men, respectively [12]. Tanning device use before the age of 35 is associated with a 75% increase in the risk for melanoma, and more than twice the risk for squamous cell carcinoma (SCC) [13]. In light of these alarming statistics, efforts are increasing to limit youth access to indoor tanning facilities. Howard County, Maryland was the first jurisdiction to ban indoor tanning for all minors under age 18 [14], and currently at least 32 states regulate tanning facility use by minors (Table 10.1). Advocating for protection of the general population from additional melanoma risk, the International Agency for Research on Cancer, a subsidiary of the WHO, encourages policymakers to consider enacting measures that limit and discourage young adults from utilizing indoor tanning facilities. While states with such policies in place did not experience a change in the prevalence of adolescent indoor tanning use from 1998 to 2004, prevalence rates increased in states without such policies [16].

While indoor tanning is not unique to the United States, the delivery method, locations of the devices and manner of marketing the service may vary from one country to another. For instance in the United Kingdom (UK), unsupervised coin-operated machines are available in salons and gyms. A UK government-commissioned study revealed that a quarter of a million British teens used tanning beds, and the "Six Cities Study of England" revealed that more than one in five children acknowledged using sunbeds in their own home, and one quarter had experienced unsupervised use in a salon or gym/leisure center [17]. When booths were supervised, only 11–37% of children said they were informed of the risks of indoor tanning [18]. Seeking to emulate France, which has established a legal minimum age for tanning (age 18) [13], British Public Health Minister Gillian Merron favors under-18 indoor tanning bans, and personally believes that voluntary action from the tanning industry is not enough to ensure protection of British youth [19, 20]. Scotland and Welsh authorities likewise are cracking down with restriction on under 18 access to both unsupervised and coin-operated salons units [21].

In South America, residents of Chile frequently use tanning salons [22]. In April, 2006 a Chilean law took effect regulating the operation of tanning salons and tanning beds. The law includes both technical specifications about tanning parlor location, as well as a requirement to obtain informed consent [22]. The worldwide use of tanning salons by underage individuals [23–25] underscores the importance of regional, national, and international efforts to protect youth from exposure to devices that emit a known carcinogen.

Table 10.1 State-specific tanning restrictions for minors[a]

State	Statute	Parental presence	Parental permission	Other
AZ	AZ Admin Code R12-1-1414 A2	–	<18, in person	Operator must limit exposure time to manufacturer's recommendation; provide eye protection
AR	AR Stat Ann § 20-27-2202	–	<18, in person	–
CA	CA Bus & Prof Code § 22706 (b) (3)&(4)	–	<18, in person; valid 12 months	Operator must limit time to manufacturer's max exposure recommendation; provide eye protection Banned <14
CT	CT Gen Stat § 19a-232	–	<16	–
DE	DE Code Ann tit 16 § 30D	–	<18, in person; valid 12 months	Banned <14; unless medically necessary
FL	FL Stat Ann § 38189 (1998)	<14	<18; agrees to wear eye protection	Operator must limit time to manufacturer's max exposure recommendation; provide eye protection
GA	GA Code Ann § 31-38-8 (1996)	–	<18, in person	Operator must provide eye protection
IL	IL Admin Code Title 77; Sec 795190 (c)	–	<18, in person	Banned <14
IN	IN Code Ann § 25-8-154-15 & 16	<16	<18, in person	Operator must limit time to administrative or manufacturer's max exposure recommendation; provide eye protection
KY	KY Rev Stat § 217922	<14	<18; agrees to wear eye protection; valid 12 months	–
LA	LA Rev Stat Ann § 40:2701 to 40:2718 (2005)	<14	<18, in person; agrees to wear eye protection	Operator must provide eye protection
ME	10-144 Dept of Human Services ch 223 12A (3)(f)	<16	<18, in person; valid for 12 months	Operator must limit time to administrative or manufacturer's max exposure recommendation; provide & require eye protection Banned <14
MD	MD Health Code Ann § 20-106	–	<18, in person	–

MA	MA Gen Laws Ann ch 111 Public Health § 211	<14	<18	Operator must limit time to administrative max exposure recommendation; provide & require eye protection
MI	MI Comp Laws Ann § 33313405	–	<18, in person	Operator must require eye protection
MN	MN Stat Ann § 325 H08	–	<16, in person	Operator must limit time to manufacturer's max exposure recommendation; provide & require eye protection
MS	Department of Health Regulations	<14	<18, in person; valid 12 months & number of tanning sessions as specified by parent	Operator must limit time to manufacturer's max exposure recommendation; provide & require eye protection
NH	NH Rev Stat Ann § tit XXX 313-A:31	<14	<18, in person; valid for 12 sessions	Banned <14; unless medically necessary
NJ	NJ Rev Stat § C 26:2D-821	–	<18	Operator must limit time to administrative max exposure recommendation; require eye protection Banned <14
NY	NY Public Health Law § 3555 S6905/A9199	–	<18, in person; valid 12 months	Operator must require eye protection Banned <16
NC	NC Gen Stat § 104E-91 15A NCAC 11 1418	–	<18	Operator must limit time to manufacturer's max exposure recommendation; provide eye protection Banned <14; unless medically necessary
ND	ND Cent Code § 23-39	<14	<18, in person; valid 12 months	Operator must limit time to manufacturer's max exposure recommendation; provide & require eye protection Banned <14; unless medically necessary
OH	OH Admin Code 4713-19-09 (B)	–	<18, in person; valid for number of tanning sessions specified by parent	Operator must limit time to manufacturer's max exposure recommendation; provide eye protection

(continued)

Table 10.1 (continued)

State	Statute	Parental presence	Parental permission	Other
OR	OR 333-119-0090 (2)	–	<18, in person	Operator must limit time to manufacturer's max exposure recommendation; provide & require eye protection
RI	Department of Health Rules & Regulations for the Registration of Tanning Facilities Part III; Sec 95	–	<18, in person	Operator must limit time to manufacturer's max exposure recommendation; provide & require eye protection
SC	SC Code Ann § ch 61, sec 106-45	–	<18, in person	Operator must provide & require eye protection
TN	TN Code Ann § 68-117-104	<14	<18, in person	Operator must limit time to manufacturer's max exposure recommendation; provide & require eye protection
TX	TX Health & Safety Code Ann § 145008	–	<18, in person	Operator must limit time to manufacturer's max exposure recommendation; provide & require eye protection Banned <16
UT	UT Code Ann § 26-15-13	–	<18, in person; valid 12 months & number of tanning sessions as specified by parent	–
VA	VA Code § 591-3103	–	<15; valid 6 months	–
WI	WI Code Ann § 25508 (9)(a)	–	–	Operator must limit time to manufacturer's max exposure recommendation; provide & require eye protection Banned <16
WY	Enrolled Act 36, 2010	<15	<18; valid 12 month	–

[a]Table adapted from National Conference of Stage Legislatures [15]

10.3 United States Federal Regulations

In the United States, the indoor tanning industry is still growing rapidly, generating over $5 billion in annual revenues and attracts over 30 million patrons a year, with over one million people, on average, using tanning beds each day [26]. The FDA is responsible for safe-guarding public health via assurance of safety and efficacy of medical devices, cosmetics, and products that emit radiation. Since tanning units used for cosmetic purposes are not considered medical devices, the FDA is currently limited to regulating their emissions; however, some are urging reclassification of the lamps in order to require FDA approval before the tanning beds are put onto the market. An FDA advisory panel convened on March 25, 2010 and recommended that tanning beds be listed as Class 2 devices, which require special assurances, such as labeling requirements or mandatory performance standards, that they will not cause harm [27]. The panel also recommended increased restrictions on the use of tanning beds for everyone under 18 years old.

Currently, manufacturers of indoor tanning equipment must adhere to FDA-mandated requirements for sunlamp specifications, posting of warning labels, and provision of appropriate eye protection [28, 29]. The sale of sunlamps and/or sunbeds that are used for cosmetic purposes are not, however, required to undergo any sort of pre-sale clearance, and most of the FDA's regulation of sunlamps occurs at the review level of product reports from manufacturers, periodic inspections of manufacturers, and infrequent FDA field-office inspections of salons.

The minimal erythemal dose (MED) is the minimal single dose of UV radiation (energy per unit area) required to produce erythema (sunburn) at a commonly sun protected site, such as the buttocks, 24 h after laboratory-controlled UV radiation exposure [30, 31]. Per FDA recommendations, UV exposure should be restricted to "no more than 0.75 minimal erythemal dose (MED), three times during the first week" of exposure, gradually increasing the exposure thereafter [29, 32]. Based on these guidelines, manufacturers and salon staff are to develop recommended exposure schedules for their clients.

In 2007 the FDA Tanning Accountability and Notification (TAN) Act (HR 4767) was signed, reflecting increased consumer interest in UV tanning risks [24]. This amendment requires that the FDA conduct consumer testing to determine if the text and positioning requirements for the warning statements on tanning devices provide adequate information regarding the risks of using tanning equipment [25]. Amidst an influx of proposed anti-tanning legislation, this action was noted by the Indoor Tanning Association (ITA) as a "wake-up call" for their national lobby to oppose banning teens from tanning beds [25]. By 2010, it seemed likely that the FDA would require clear labeling on the devices definitively stating that UV radiation causes skin cancer. This was a change from prior warnings that stated "repeated exposure may cause premature aging of the skin and skin cancer [33]."

The sun emits UV radiation of three different wavelengths: UVA (320–400 nm), UVB (290–320 nm), and UVC (200–290 m). UVC emitted from the sun is absorbed by the atmosphere's ozone layer and plays little to no role in tanning derived from natural sunlight. Longer wavelengths of UV light penetrate the skin more deeply.

UVB is responsible for burning, but UVA penetrates through the epidermis and into the deeper levels of the dermis. While there are limitations on the amount of UVC that can be emitted by indoor tanning beds, the FDA does not regulate the proportion of UVA and UVB emitted [34]. Thus, the proportions may vary from salon to salon and bed to bed, making it difficult for consumers to predict the results of any particular indoor tanning session. Furthermore, adherence to proposed tanning schedules is not regulated. One study randomly surveyed patron exposure records during routine state inspections of 50 North Carolina indoor tanning facilities and revealed that 95% of clients exceeded the recommended limits and that 33% began tanning at the maximum recommended doses [34]. Other studies of New York and California-based tanning salons also revealed overwhelming noncompliance with FDA guidelines [35, 36].

In 2004, the AAD, American Society of Photobiology, and the FDA identified and mutually agreed upon recommendations and courses of action of which the highest priority was restriction of access to minors [31]. In recognizing the hazards of exposure to ultraviolet radiation, the FDA aims to reduce UV burden to the public, recommending that people not use sunlamps. The FDA, however, has so far failed to institute a full-fledged ban on indoor tanning [31]. In 2009, the AAD encouraged the FDA to prohibit claims for beneficial effects from tanning as this practice is deceptive [37].

10.4 Indoor Tanning Advertisements

Indoor tanning businesses target adolescents by advertising in local newspapers that target high school and college students [38, 39]. In addition to targeting specific populations, indoor tanning advertisement strategies promote UV exposure as "necessary" for Vitamin D synthesis; appeal to the desire for social acceptance; and describe mood altering effects (e.g. decreased anxiety, increased relaxation) [40]. These ads offer discounts and promotions for unlimited tans, while neglecting to mention the associated risks of frequent tanning [41]. Kwon and colleagues conducted two studies: the first study monitored tanning facility advertisements in 24 San Diego County newspapers over 4 months, while the second study conducted telephone interviews of 60 San Diego County tanning facilities to assess compliance with recommended exposure schedules [36]. Up to three-fourths of indoor tanning print ads offered "unlimited" tanning packages, and all facilities examined offered unlimited tanning for a flat monthly fee of about $29 [36]. Legislation aimed at limiting or prohibiting tanning advertisements targeting minors is one proposed method to reduce indoor tanning [38].

In the 1980s, the indoor tanning industry began using data for the carcinogenicity of UVB rays to promote UVB-free and "99% pure UVA" tanning beds [42]. Advertisers continue to promote their equipment as offering the "ideal" proportion of UVA to UVB rays, while neglecting to mention that UVA rays also have carcinogenic potential [42, 43]. Included in this pseudo "harm reduction" tactic, advertisers also promote the "health benefits" of tanning, capitalizing on the controversies surrounding vitamin D and sun exposure [44]. The ITA has aired a series of commercials advocating that tanning devices are "a great way to get a healthy dose of vitamin D [45]," while failing to mention that 15–45 min per day of outdoor sun exposure to the face and hands alone is sufficient to stimulate adequate vitamin D [44] and that additional exposure does not further increase vitamin D levels [46]. See Chap. 8 by Mendese and Gilchrest for a much more detailed discussion of the vitamin D controversy.

The lack of universal regulations or restrictions relating to indoor tanning advertising facilitates questionable marketing claims and aggressive targeting of minors. Various states now have in place, or are in the process of adopting, regulations of such tactics. For example, the Texas' Tanning Facility Regulation Act states: "a tanning facility operator may not claim or distribute promotional materials that claim using a tanning device is safe or free from risk or that using a tanning device will result in medical or health benefits." As a result of this Act, the Texas Attorney General charged Darque Tan™ tanning salon operators with conducting unlawful marketing campaigns when their company aired commercials and billboards across Austin, TX in violation of this code. Darque Tan™ was issued a warning letter, and injunctions sought to halt the misleading practices and return money paid by customers misled by the marketing campaigns' inaccurate claims [47].

10.5 Compliance with Regulations

The sheer number and variety of new legislation in 2010 will make efforts to document non-compliance, and variability in business practices a challenge to enforce [48]. Of those tanning facilities assessed during the late 1980s and 1990s in North Carolina (n=50), New York (n=20), and San Diego (n=60), either by operator survey or random patron record survey, greater than 95% of customers were found to have exceeded exposure limits [42, 44, 45, 50]. Out of 28 states with tanning legislation, less than 50% of the cities in each state gave citations for indoor tanning facility violation, 32% did not perform inspections, and 32% performed inspections less than once a year as of 2007 [10].

In addition to state regulations, non-compliance with federal regulations of indoor tanning is likewise problematic. Of those tanning facilities investigated using research personnel posing as customers (confederates) in 116 heavily populated US cities, the average rate of compliance with FDA-recommended session-per-first-week frequency was only 11% [50]. Confederates were even allowed to tan everyday by a significant proportion of staff. Consistent with the findings of numerous smaller studies [35, 43, 51], the authors suggest that these data underscore the value of regulation over a recommendation [50]. Lending further support to the need for FDA regulation, one study demonstrated that during 1985–2006, 36% of UV radiation-attributable indoor tanning injury reports filed with the FDA and Consumer Product Safety Commission occurred in the setting of non-compliance with FDA recommendations [54].

10.5.1 Education of Tanning Facility Operators and Employees

Tanning facility operator/employee education may help reduce injury and increase implementation of indoor tanning regulations. Culley and colleagues found that approximately 96% of tanning bed operators in the US incorrectly replied affirmatively to the question "Does a base tan protect me from a burn?"[43] In another study, 46% of indoor tanning operators either claimed they "did not know" about or misinformed patrons of the risk of skin cancer associated with tanning device usage [55]. Studies in Chile, Poland, and Argentina have similarly demonstrated inadequate indoor tanning facility employee knowledge [54–56]. These studies were unable to address if employees feigned ignorance or truly were uneducated regarding the risks of indoor tanning. Of note, tanning centers in Poland were found to have high safety standards and were more likely to have employees with more knowledge, thus, reinforcing the need to have qualified tanning facility operators [56]. Proponents of educating indoor tanning facility staff also advocate that staff should be trained to conduct a focused clinical history and brief physical exam to identify those potential clients at an increased risk for skin cancer who should avoid UV exposure [57].

A required license to deliver indoor tanning sessions could help fund tanning salon inspection efforts, thus helping to ensure adherence to regulations. Licensed tanning bed operators should be trained regarding the risks associated with indoor tanning.

10.6 The Role of Parental Consent and Education in Youth Access to Indoor Tanning

In addition to prohibiting tanning bed use for those under the age of 16.5, Texas recently instituted legislation mandating that tanners under the age of 18 must have written consent from a parent to use an indoor tanning facility [58]. Likewise, under current Maryland law, minors are only allowed tanning device use after obtaining parental or legal guardian consent [49, 59]. Where written parental consent is required, teens may forge their parent's signature [60]. This was the case in Indiana, when one teen who, after confronted by her mother, admitted to forging her parent's signature in order to get a spray tan [61]. Opponents of required parental consent for tanning also maintain that it threatens parental rights as summarized by Texas Congressman Ralph Hall: "I don't think you need any legislation either in the state or in the federal government telling mothers and dads how to treat their 16 year old [58]." This *laissez-faire* attitude towards parental consent for cosmetic indoor tanning is in contrast to the approach towards medical UV radiation. All states, except Louisiana, require parental consent for medical UV treatments in patients less than 18 years of age [15, 53].

Parents play a critical role in shaping adolescent tanning behaviors [16, 63–65]. Studies not only support that parental tanning bed use is a significant predictor of their children's subsequent use of tanning devices [66, 67] but also suggest that children use tanning devices earlier than parents, equating to more years of cumulative UV damage [67]. Adolescents are more likely to tan indoors if their parents allow them to do so [67, 68]. Because parental role-modeling is a critical factor in the prevention of substance abuse, tobacco use, and antisocial behavior [69], it follows, then, that parents may play a vital role in reducing skin cancer risk behaviors such as early and repetitive tanning device use in teens and pre-teens.

Parents serve both as role-models and gatekeepers of tanning behavior in their children and adolescents. As efforts to require parental consent gain traction, ensuring that parents understand the risks of indoor tanning will help ensure that truly informed consent is obtained. One study revealed that a majority of parents were under the impression that a tan from a tanning device would be protective against future sunburns [67]. From 1998 to 2004, indoor tanning significantly increased among surveyed parents of US adolescents, with 23% of parents practicing indoor tanning [65]. Interestingly, teens were more likely to have tanned indoors within the preceding 12 months if their parents concurred that indoor tanning might cause skin cancer compared to teens whose parents did not agree [68]. Although counterintuitive, parents with higher education and higher socioeconomic status were more likely to have ever used indoor tanning devices and less likely to have talked about tanning device use with their children [67], but there was no significant association between parents' knowledge and children's usage of tanning devices in this study. These parental attitudes and beliefs towards tanning may shape their decision to allow or disallow their children to tan indoors, underscoring the concomitant need for education efforts targeting parental knowledge and attitudes regarding indoor tanning in order for parental consent measures to be meaningful.

10.7 Lessons Learned from Tobacco Control

Tobacco products, like indoor-tanning devices, are associated with serious health risks, place emphasis on social appeal in advertizing, and are commonly-used by adolescents. Over the last 40 years, effective tobacco control efforts in the United States have resulted in substantial declines in tobacco use. Effective policies limiting youth access to tobacco products therefore are useful models for tanning legislation strategies [70]. Youth, perhaps because of lower incomes, are particularly cost-sensitive, and tobacco taxation is the single most effective intervention for reducing its demand [71, 72]. Tobacco control campaigns effectively use sports icons and teen idols, as well as humor, to promote no-use messages. Such strategies, while costly, could also be used in changing youth indoor tanning behavior.

10.7.1 Excise Tax on Indoor Tanning

Proponents of taxing customers for tanning sessions suggest that this could decrease demand, while simultaneously creating a pool of funding for enforcement and public health efforts towards education for skin cancer prevention, [73] as has been the case with tobacco taxation. On March 23, 2010, President Obama signed into law the Health Care Reform H.R. 3590, the Patient Protection and Affordable Care Act, which includes a federal excise tax on indoor tanning services: "a tax equal to 10% of the amount paid for such service [74]". Furthermore, the federal law states that: "The tax imposed shall be paid by the individual on whom the service is performed." While the intention of the law is to have the increased cost serve as a deterrent to indoor tanning, the owners of tanning establishments have stated that they will pay the tax themselves rather than pass the tax on to the customer. The law does allow the tax to be paid by the person who provides the service in the event that the tax is not collected at the time the service is provided. The effective date of this tax is July 1, 2010. The tax could be insufficient to deter regular seasonal tanners, and is, therefore, more likely to deter event tanners, who are often first time users with limited income (i.e. teenagers) [75]. If unlimited tanning sessions are abolished, then taxation of single sessions may become a more effective deterrent to indoor tanning. Given these points of contention, a necessary part of instituting the indoor tanning tax will be to incrementally increase taxation rates to a price-point that serves as a barrier to indoor tanning.

10.7.2 Restriction of Indoor Tanning Advertising Directed to Youth

Advertising by the UV tanning industry directs its pro-tanning messages to youth and often conveys an inaccurate message regarding the perceived health benefits of having a tan. As in tobacco control, laws prohibiting youth-oriented advertising could aid in quelling these perceptions and protecting youth from misinformation. The Indoor Tanning Association's statements of "no increased risk of skin cancer from tanning beds" similarly fuel plaintiffs aimed at fraud claims [76].

The FTC, which regulates advertising and prohibits the use of false and deceptive statements, is an appropriate venue to press the issue of deceptive advertising. After an almost 2-year effort by the AAD, the FTC charged the tanning industry with purporting false health and safety claims about indoor tanning. On January 26, 2010, the ITA agreed to a settlement that bars it from further deception. In the future, ads that make claims about the health and safety benefits are required to clearly and prominently make this disclosure: **"NOTICE:** Exposure to ultraviolet radiation may increase the likelihood of developing skin cancer and can cause serious eye injury." Ads that make claims about the need for indoor tanning to generate vitamin D must clearly and prominently make this disclosure: **"NOTICE:** You do not need to become tan for your skin to make vitamin D. Exposure to ultraviolet radiation may increase the likelihood of developing skin cancer and can cause serious eye injury."

10.8 Conclusion

The overarching purpose of limiting youth access to indoor tanning is harm reduction. The International Commission on Non-Ionizing Radiation Protection and the WHO recommend harm reduction policies where artificial tanning devices are available, [7, 77]. These regulations and policy recommendations are grounded in years of epidemiologic and behavioral research by many scientists. However, scientific evidence rarely suffices to change market behavior without a policy to marshal the resources of professional and lay organizations to advocate for change. Tobacco control programs presaged indoor tanning control; thus, the progress and opportunities of tobacco control provide a framework for the next steps with the indoor tanning effort. As illustrated in the timeline presented in Table 10.2, over the decades, there have been many individual "champions" who helped the cause to endure. Many

Table 10.2 Timeline of major US regulatory policy and taxation of indoor tanning

Date	Organization	Action
Nov 25, 1987	AAD petitioned the US Surgeon General	C. Everett Koop, MD From the Surgeon General: Tanning and Tanning Salons can be hazardous to your health
June 1988	AAD calls for regulation of tanning parlors at the local level	Community action program with model legislation
June 1990	Illinois State Medical Society Resolution 157 (A-90) Adopted by House of Delegates of AMA to strengthen state laws to support tanning parlors regulation by state department of public health to make the consumer as informed and safe as possible	AMA: distributes State Health Legislation Report: "Tanning Parlor Regulation: A paradigm for physician involvement in public health issues" to each state department of health
Dec 1994	ASDS Resolution 217: restrict the non-medical use of tanning devices by banning the interstate shipment of tanning parlor equipment	Meeting with FDA Jan 20, 1995 by representatives of ASDS, AAD, SID to develop additional restrictions
April 2003	Comparison of UV tanning access laws to tobacco youth access laws	Provides direction for future regulation and legislation
June 2003	Tanning symposium with FDA leaders	Restrict youth access
Sept 2003	Editorial suggests taxation of each tanning session	Call to action

(continued)

Table 10.2 (continued)

Date	Organization	Action
June 2006	AMA resolution	Model state legislation to prohibit sale to those under 18 years of age
Jan 2010	AADA	FTC charged the tanning industry with making false health and safety claims about indoor tanning
Dec 2009	AADA presents taxation of indoor tanning to Senator Harry Reid	March 23,2010 H.R. 3590 signed 10% federal excise tax effective July 1, 2010
March 25, 2010	FDA Panel	Advised the FDA to list tanning beds as Class 2 devices

AAD American Academy of Dermatology, *AADA* American Academy of Dermatology Association, *ASDS* American Society for Dermatologic Surgery, *ASP* American Society for Photobiology, *SID* Society for Dermatology

dermatologists, as well as skin cancer patients and their families, have volunteered their time, money, and expertise to advocate for restricting indoor tanning with local, state, and national government. Much has been accomplished to reduce the harm of indoor tanning, but much remains to be done.

References

1. Spencer JM, Amonette RA (1995) Indoor tanning: risks, benefits, and future trends. J Am Acad Dermatol 33(2 Pt 1):288–298
2. Meadows M (2003) Don't be in the dark about tanning. FDA Consum 37(6):16–17
3. (2005) Indoor tanning: unexpected dangers. Consum Rep 70(2):30–33
4. Hilton L (2002) The darker side of tanning. Dermatology insights. American Academy of Dermatology, Schaumburg, p 11
5. Centers for Disease Control (2009) Cancer: protect yourself from the sun. http://www.cdc.gov/cancer/skin/basic_info/prevention.htm. Accessed 29 Dec 2009
6. American Academy of Dermatology (2009) Dermatology association calls for tighter regulations on indoor tanning. http://www.aadassociation.org/policy/tanning.html. Accessed 29 Dec 2009
7. World Health Organization, Geneva (2003) Artifical tanning sunbeds: risks and guidance. http://www.who.int/uv/publications/sunbedpubl/en/. Accessed 9 Jan 2011
8. Council on Scientific Affairs (1989) Harmful effects of ultraviolet radiation. JAMA 262(3):380–384
9. Balk SJ, Geller AC (2008) Teenagers and artificial tanning. Pediatrics 121(5):1040–1042
10. Mayer JA, Hoerster KD, Pichon LC et al (2008) Enforcement of state indoor tanning laws in the United States. Prev Chronic Dis 5(4):A125
11. Wilson CB (2010) Tanning parlor ultraviolet radiation. Reports of Board of trustees, American Medical Association. Volume 2010. http://www.ama-assn.org/resources/doc/hod/i-obbot.pdf. Accessed 30 Nov 2010
12. Autier P (2004) Perspectives in melanoma prevention: the case of sunbeds. Eur J Cancer 40(16):2367–2376

13. Young AR (2004) Tanning devices–fast track to skin cancer? Pigment Cell Res 17(1):2–9
14. Cavanaugh CC (2009) Howard County bans youth tanning. Health NY 130(2):18–19
15. National Conference of State Legislation (2010). Tanning restrictions for minors, a state-by-state comparison. http://www.ncsl.org/default.aspx?tabid=14394. Accessed 12 Jan 2011
16. Cokkinides V, Weinstock M, Lazovich D et al (2009) Indoor tanning use among adolescents in the US, 1998 to 2004. Cancer 115(1):190–198
17. Thomson CS, Woolnough S, Wickenden M et al (2010) Sunbed use in children aged 11–17 in England: face to face quota sampling surveys in the National Prevalence Study and Six Cities Study. BMJ 340:c877
18. Thomson CS, Twelves C (2009) Legislation is needed to stop children using sunbeds. BMJ 339:b4643
19. BBC News (2009) Cardiff MP says salon tan ban possible before election. http://news.bbc.co.uk/2/hi/uk_news/wales/south_east/8417149.stm. Accessed 29 Dec 2010
20. BBC (2009) New sunbed controls to be urged. http://news.bbc.co.uk/2/hi/health/8104250.stm. Accessed 29 Dec 2010
21. BBC (2009) Under-18s face sunbed salons ban. http://news.bbc.co.uk/2/hi/uk_news/scotland/8385432.stm. Accessed 29 Dec 2010
22. Salud MD (2007) Reglamento de solariums o cams solares. Diario Oficial 706
23. US Food and Drug Administration (2009) Tanning: radiation-emitting products. Radiation-emitting products: sunlamps and sunlamp products (tanning beds/booths). http://www.fda.gov/radiation-emittingproducts/radiationemittingproductsandprocedures/tanning/default.htm. Accessed 21 Mar 2010
24. US Food and Drug Administration (2007) Tanning Accountability and Notification (TAN) Act. 110–085
25. Reykdal P, Smith D (2007) TAN act signed into law. Looking fit. http://www.lookingfit.com/hotnews/7ah2105842.html#. Accessed 29 Dec 2010
26. Bizzozero J. The State of The Industry Report 2008. http://www.lookingfit.com/articles/state-of-the-industry-2008-fact-book-2009.html. Accessed 26 Mar 2010
27. DeNoon D (2010) FDA panel: restrict tanning beds; children and teens would need parental consent in proposed regulation. Accessed 3 May 2010
28. United States Department of Health and Human Services, Food and Drug Administration (1985) Sunlamp products performance standard; final rule. United States Department of Health and Human Services, Food and Drug Administration, Rockville
29. United States Department of Health and Human Services (1986) Policy on maximum timer interval and exposure for sunlamp products. United States Department of Health and Human Services, Rockville
30. Whitmore SE, Morison WL, Potten CS et al (2001) Tanning salon exposure and molecular alterations. J Am Acad Dermatol 44(5):775–780
31. Lim HW, Cyr WH, DeFabo E et al (2004) Scientific and regulatory issues related to indoor tanning. J Am Acad Dermatol 51(5):781–784
32. United States Department of Health and Human Services (1988) Quality control guide for sunlamp products. United States Department of Health and Human Services, Rockville
33. Dooren J (2010) FDA likely to tighten tanning bed industry regulations. Wall St J. http://online.wsj.com/article. Accessed 23 Aug 2011
34. Hornung RL, Magee KH, Lee WJ et al (2003) Tanning facility use: are we exceeding Food and Drug Administration limits? J Am Acad Dermatol 49(4):655–661
35. Fairchild AL, Gemson DH (1992) Safety information provided to customers of New York City suntanning salons. Am J Prev Med 8(6):381–383
36. Kwon HT, Mayer JA, Walker KK et al (2002) Promotion of frequent tanning sessions by indoor tanning facilities: two studies. J Am Acad Dermatol 46(5):700–705
37. Riegel D (2009) http://www.vertiy.fda.gov/search97cgi/s97_cgi.exe. Accessed 30 Dec 2009
38. Lazovich D, Forster J (2005) Indoor tanning by adolescents: prevalence, practices and policies. Eur J Cancer 41(1):20–27
39. Freeman S, Francis S, Lundahl K et al (2006) UV tanning advertisements in high school newspapers. Arch Dermatol 142(4):460–462

40. Greenman J, Jones DA (2010) Comparison of advertising strategies between the indoor tanning and tobacco industries. J Am Acad Dermatol 62(4):685, e681–618
41. Hester EJ, Heilig LF, D'Ambrosia R et al (2005) Compliance with youth access regulations for indoor UV tanning. Arch Dermatol 141(8):959–962
42. Sun Italia advertisement. http://www.extremetanning.com/tanning-beds.html. Accessed 29 Dec 2010
43. Culley CA, Mayer JA, Eckhardt L et al (2001) Compliance with federal and state legislation by indoor tanning facilities in San Diego. J Am Acad Dermatol 44(1):53–60
44. Reichrath J (2006) The challenge resulting from positive and negative effects of sunlight: how much solar UV exposure is appropriate to balance between risks of vitamin D deficiency and skin cancer? Prog Biophys Mol Biol 92(1):9–16
45. Hines JR (2010) http://www.youtube.com/watch?v=MWIVeuBHrcQ. Accessed 26 Dec 2010
46. Holick MF, MacLaughlin JA, Clark MB et al (1980) Photosynthesis of previtamin D3 in human skin and the physiologic consequences. Science 210(4466):203–205
47. Strickland J, Kelley T (2008) Attorney general charges tanning salon operators with conducting unlawful marketing campaign. Austin News: Austin:Attorney General of Texas, Greg Abott. Nov 2008
48. Paulson A (2010) New laws for 2010: no texting, trans fats, or tanning bed. The Christian Science Monitor: ABC News
49. Schultz D (2009) No more teen tanning in Howard County, Maryland. http://wamu.org/news/09/11/11.php#30115. Accessed 14 Jan 2010
50. Pichon LC, Mayer JA, Hoerster KD et al (2009) Youth access to artificial UV radiation exposure: practices of 3647 US indoor tanning facilities. Arch Dermatol 145(9):997–1002
51. Hurd AL, Mayer JA, Woodruff SI et al (2006) Comparing two methods of measuring legislation compliance among indoor tanning facilities. J Am Acad Dermatol 54(3):433–439
52. Dowdy JC, Sayre RM, Shepherd JG (2009) Indoor tanning injuries: an evaluation of FDA adverse event reporting data. Photodermatol Photoimmunol Photomed 25(4):216–220
53. Heilig LF, D'Ambrosia R, Drake AL et al (2005) A case for informed consent? Indoor UV tanning facility operator's provision of health risks information (United States). Cancer Causes Control 16(5):557–560
54. Chouela E, Pellerano G, Bessone A et al (1999) Sunbed use in Buenos Aires, Argentina. Photodermatol Photoimmunol Photomed 15(3–4):100–103
55. Salomone C, Majerson D, Molgo M et al (2009) Tanning salons in Santiago, Chile: the knowledge of the staff in charge and the quality of information provided to potential clients before and after a new regulatory law. Photodermatol Photoimmunol Photomed 25(2):86–89
56. Szepietowski JC, Nowicka D, Soter K et al (2002) Tanning salons in southwest Poland: a survey of safety standards and professional knowledge of the staff. Photodermatol Photoimmunol Photomed 18(4):179–182
57. International Commission on Non-Ionizing Radiation Protection Statement (2003) Health issues of ultraviolet tanning appliances used for cosmetic purposes. Health Phys 84(1):119–127
58. Sahr H (2010) New law for Texas teen tanners. http://www.kxii.com/home/headlines/80519712.html. Accessed 14 Mar 2010
59. PR Newswire UBM, Annapolis MD (2010) Maryland legislation introduced to prohibit indoor tanning for minors. http://www.prnewswire.com/news-releases/maryland-legislation-introduced-to-prohibit-indoor-tanning-for-minors-87237602.html. Accessed 26 Dec 2010
60. Schneider M (2005) Legislating teens' indoor tanning, state by state: restrictions on minors are lacking and vary widely; the AMA urges federal prohibitions and warnings. Health Care Ind 15(4):112–115
61. Loncich J (2009) Major controversy after a teenager gets a spray tan. Indianapolis. Fox 59 WXIN. Aired on 18 Dec 2009
62. Dellavalle RP, Parker ER, Cersonsky N et al (2003) Youth access laws: in the dark at the tanning parlor? Arch Dermatol 139(4):443–448

63. O'Riordan DL, Geller AC, Brooks DR et al (2003) Sunburn reduction through parental role modeling and sunscreen vigilance. J Pediatr 142(1):67–72
64. Stryker JE, Lazovich D, Forster JL et al (2004) Maternal/female caregiver influences on adolescent indoor tanning. J Adolesc Health 35(6):528, e521–529
65. Bandi P, Cokkinides VE, Weinstock MA et al (2010) Sunburns, sun protection and indoor tanning behaviors, and attitudes regarding sun protection benefits and tan appeal among parents of U.S. adolescents-1998 compared to 2004. Pediatr Dermatol 27(1):9–18
66. Cokkinides VE, Weinstock MA, O'Connell MC et al (2002) Use of indoor tanning sunlamps by US youth, ages 11–18 years, and by their parent or guardian caregivers: prevalence and correlates. Pediatrics 109(6):1124–1130
67. Magee KH, Poorsattar S, Seidel KD et al (2007) Tanning device usage: what are parents thinking? Pediatr Dermatol 24(3):216–221
68. Hoerster KD, Mayer JA, Woodruff SI et al (2007) The influence of parents and peers on adolescent indoor tanning behavior: findings from a multi-city sample. J Am Acad Dermatol 57(6):990–997
69. Dishion TJ, McMahon RJ (1998) Parental monitoring and the prevention of child and adolescent problem behavior: a conceptual and empirical formulation. Clin Child Fam Psychol Rev 1(1):61–75
70. Loh AY (2008) Are artificial tans the new cigarette? How plaintiffs can use the lessons of tobacco litigation in bringing claims against the indoor tanning industry. Mich Law Rev 107(2):365–390
71. Chaloupka FJ, Cummings KM, Morley CP et al (2002) Tax, price and cigarette smoking: evidence from the tobacco documents and implications for tobacco company marketing strategies. Tob Control. 2002;11 Suppl 1:I62–I72
72. Jha P, Chaloupka FJ. The economics of global tobacco control. BMJ. 2000;321(7257):358–361
73. Dellavalle RP, Schilling LM, Chen AK et al (2003) Teenagers in the UV tanning booth? Tax the tan. Arch Pediatr Adolesc Med 157(9):845–846
74. CNN (2010) Obama signs health care bill; Senate takes up house changes. http://articles.cnn.com/2010-03-23/politics/health.care.main_1_health-care-obama-and-democratic-leaders-health-insurance?_s=PM:POLITICS. Accessed Dec 2010
75. Hillhouse J, Turrisi R, Shields AL (2007) Patterns of indoor tanning use: implications for clinical interventions. Arch Dermatol 143(12):1530–1535
76. Indoor tanning association settles FTC charge that it deceived consumers about skin cancer risks from tanning. FTC file No 0823159. Available from: http://www.ftc.gov/opa/2010/01/tanning.shtm
77. International Commission on Non-Ionizing Radiation Protection (ICNIRP). Health issues of ultraviolet tanning appliances used for cosmetic purposes. Health Phys 2003;84(1):119–27

Chapter 11
Sunless Tanning

Sherry Pagoto

Abstract Sunless tanning, also referred to as self-tanning, fake tanning, and/or chemical tanning, includes agents that produce the appearance of a tan without exposure to ultraviolet radiation (UVR). Sunless tanning agents include products containing dihydroxyacetone (DHA), tanning pills, and cosmetic bronzers. DHA-based products, the fastest growing skin care product on the market, have annual use rates estimated in the last several years at 11% in a nationally representative sample, 24% in college students, and 46% in beach visitors. In this chapter, the safety, risks, clinical applications, use rates, user characteristics, and intervention research regarding DHA-based sunless tanning products are reviewed. At this time, no known risks have been associated with the application of DHA-based sunless tanning products to the skin. However, users should be educated about the need for sun protection while using sunless tanning. Skin cancer prevention intervention research reveals some benefits but no harms to promoting sunless tanning as an alternative to UVR tanning among tanners. Sunless tanning is a safe alternative to UVR tanning and should be recommended to UVR tanners who are primarily motivated by the desire to tan to improve physical appearance. Sunless tanners may increase in popularity as products continue to improve in quality and include UVR protection. Sunless tanning products that are currently in development and create a tan via melanin production stimulation, rather than DHA, are also discussed.

Keywords Sunless tanning • Chemical tanning • Artificial tanning • Self-tanning • Fake tanning • Bronzer • Sherry Pagoto • Skin cancer • Melanoma • Dihydroxyacetone • DHA • Indoor tanning • Melanin • Maillard reaction • Eva Wittgenstein

---

S. Pagoto (✉)
Department of Medicine, Division of Preventive and Behavioral Medicine,
University of Massachusetts Medical School, 55 Lake Street North, Worcester, MA 01655, USA
e-mail: Sherry.pagoto@umassmed.edu

C.J. Heckman and S.L. Manne (eds.), *Shedding Light on Indoor Tanning*,
DOI 10.1007/978-94-007-2048-0_11, © Springer Science+Business Media B.V. 2012

Abbreviations

DHA	Dihydroxyacetone
FDA	Food and Drug Administration
MCR-1	Melanocortonin receptor-1
MSH	Melanin-stimulating hormone
SPF	Sun protection factor
UVR	Ultraviolet radiation

11.1 Definition and History

Sunless tanning, also referred to as self-tanning, fake tanning, and/or chemical tanning, includes any agent that produces the appearance of a tan without direct exposure to ultraviolet radiation (UVR). Sunless tanning agents include products containing dihydroxyacetone (DHA), tanning pills, and cosmetic bronzers. All three of these agents will be defined in this section, however this chapter will largely focus on DHA-containing agents, given that these are the most commonly used form of sunless tanning and have been the focus of the vast majority of the existing literature.

11.1.1 Dihydroxyacetone-Containing Agents

The most commonly used form of sunless tanning is DHA-based products, which are applied directly to the skin. DHA is a colorless vegetable-derived sugar that interacts with dead surface cells in the epidermis, thereby staining the skin a tan color [1]. Eva Wittgenstein, a scientist at the University of Cincinnati, accidentally discovered the skin-staining effect of DHA in the 1950s while researching DHA as an oral agent for children with glycogen storage disease [2]. When DHA made skin contact she observed that a browning effect developed several hours later. This effect, known as a Maillard reaction [3] is a chemical reaction between DHA and amino acids within skin proteins, creating pigments called melanoids. This occurs in the stratum corteum [4], the outermost layer of the skin, which is primarily composed of dead skin cells. The darkening of the skin develops over 2–4 h and cannot be removed with water or soap but only via pigment loss resulting from sloughing of the skin cells, which occurs over the course of a few days [5].

Following the initial discovery of the skin-staining effect, the cosmetic industry developed a series of DHA-containing over-the-counter products for consumers desiring a quick tan. Products come in the form of creams, foams, sprays, and wipes, to be applied directly to the skin. The first cosmetic product "QT" which stood for "Quick Tan" was put forth in the 1960s by Coppertone but failed to thrive in the marketplace because of complaints of bad odor and that it turned the skin an undesirable orange hue. In 1973 the U.S. Food and Drug Administration (FDA) formally approved DHA as a color additive for use in cosmetics [6], and shortly thereafter the next generation of DHA-containing sunless tanning cosmetic products appeared in the marketplace, improving upon the odor and color problems. According to the National Association of Chain Drug Stores Sun Safety Alliance, sunless tanning products are now the fastest growing skin care product on the market [7]. Some problems still remain with some sunless tanning products, including color, odor, and misapplication. Misapplication can produce poor results, including uneven coloring, streaking, and/or color in unwanted areas, which is a challenge because the product cannot simply be washed off.

Tanning salons have also responded to the demand for sunless alternatives and the challenges of misapplication by adding sunless tanning services (e.g., spray booths and airbrushing) to their businesses. Spray booths are shower-sized stalls equipped with nozzles that spray a mist of DHA-containing sunless tanning product onto the user's body. Thus, the user is spared the task of application and the risk of misapplication. The first spray tan booth was introduced in 1999. More recently, airbrush tanning in which the tanning product is applied by a trained cosmetician via a handheld airbrush device was introduced as a service. By 2003, 43% of tanning businesses reported offering some form of sunless tanning service [8], a number that has likely continued to rise. Spray booth and airbrushing provide a single application that is approximately four to six times more expensive than a bottle of sunless tanning lotion, which provides multiple applications, but require relatively minimal effort from the user. However, because the risk of misapplication is lower, a booth or airbrush user can start with a darker color rather than gradually increase the color intensity with repeated applications as self-applied users often do.

11.1.2 Tanning Pills

Canthaxanthin, a carotenoid found in plants and animal, is most often used as a food additive to color foods, but when ingested in large quantities in pill form can tint the skin an orange color by attaching to fat molecules underneath the skin. Although carotenoids are found in many foods, the quantity used in tanning pills is far higher, which can lead to serious side effects. Such side effects include canthaxanthin-induced retinopathy which results when yellow canthaxanthin crystals deposit in the retina of the eye [9], liver injury, and hives Canthaxanthin is not FDA-approved for use as an oral tanning agent. The FDA states that it is currently illegal to sell such products in the U.S., and imported products are subject to detention [9], however, such products are widely available for purchase on the Internet.

11.1.3 Cosmetic Bronzers

Cosmetic companies have put forth make-up bronzers which come in the form of powders, creams, and sprays to be applied directly to the skin to create a tanned appearance. Such products are similar to other forms of make-up in that they do not penetrate the skin surface and can be easily washed off with soap and water. Products include liquid facial makeup, powders for facial and/or body application, and sprays for body application (e.g., spray-on pantyhose). Bronzers are often added to DHA-containing products to create an immediate tan since DHA take several hours to develop. The bronzer also helps reduce misapplication by helping users to see where they have applied the product.

11.2 Safety and Adverse Effects of DHA-Based Sunless Tanners

The American Academy of Dermatology [10], the Skin Cancer Foundation [11], and the American Cancer Society [12] advocate use of DHA-containing sunless tanning as a safe alternative to UV tanning. DHA is both nontoxic [5] and hypoallergenic [13]. Contact dermatitis is rarely reported, however one study showed that frequent and prolonged exposure to DHA caused severe contact dermatitis in Mexican hairless dogs [14], although no evidence of dermititis was observed following frequent and prolonged exposure in mice [15]. The literature on safety issues and risks of use addresses the issues of photoprotection, skin cancer risk, non-FDA approved exposures, and effect on melanocytic nevi or moles.

11.2.1 Photoprotection

Melanin is a dark pigment produced by melanocytes in the skin. Melanin production is activated by UVR thereby creating a tan, which affords subsequent UVR protection to the skin, although this can be at the expense of DNA damage. Tanners often refer to this protection as a "base tan," and some intentionally build up melanin levels in the skin for the purpose of reducing vulnerability to painful sunburns. As discussed previously, DHA "tans" the skin simply by staining dead skin cells in the stratum corneum, but it in no way affects melanin production. Very little sun protection is afforded by DHA-containing sunless tanning formulations. Sun protection occurs via melanoids, which are brownish polymers produced by interactions between DHA and amino acids in the stratum corneum. Melanoids have a UVR-A sun protection factor of 2–4 but provide no appreciable UVR-B protection [16]. The sun protection factor gradually declines as dead skin cells slough off over the course of a few days. Because DHA has no impact on melanin production, the amount of sun protection is not comparable to melanin, even when the resulting color is comparable to a UVR tan. One study showed that DHA-treated skin produced 180% more free radicals in response to UV exposure compared to untreated skin, and the authors concluded that additional UV protection is needed for DHA-treated skin because photoaging is actually accelerated [17]. In general, evidence does not support any significant UVR protection effect of DHA.

That DHA creates the appearance of a tan without the comparable protection of a UVR-induced tan (i.e., via melanin) could lead people to inadvertently expect more sun protection than they are actually receiving, and this false sense of security could lead to greater intentional sun exposure. Some companies add sunscreen with higher sun protection factor (SPF; 15 or greater) to their sunless tanning products, but the sun protection effect does not last as long as the tanning effect, and sunscreen reapplication is required at much shorter intervals than are necessary for sunless tanner reapplication. The same is true for other cosmetic products that include sunscreen (e.g., liquid and powder makeup), in which case supplemental sunscreen and sunscreen reapplication is likely necessary for adequate UVR protection. The FDA has mandated that sunless tanning products containing no added sunscreen include a warning label that clearly states that the product is not a source of sun protection [18]. In spite of FDA labeling regulations, 33% of survey respondents in a 2006 study reported that they believe that sunless tanning products provide sun protection [19]. This perception could be in part driven by the fact that some products incorporate sunscreen. Health messages may be needed for users of DHA-containing sunless tanning products to correct misconceptions about the fact that these products are not an adequate source of sun protection.

11.2.2 Skin Cancer Risk

Whether DHA influences skin cancer risk has not been well-studied. One animal study showed treating mice with 20% concentration of DHA delayed the appearance of tumors following exposure to broad spectrum UVR, compared to mice not treated with DHA [20]. The application of 5% DHA, which is more similar to the concentration contained by over-the-counter products (1–15%) did not provide similar protection against tumor development. The investigators speculated that the 20% DHA exposure may have reduced the amount of UVR penetrating the epidermis, thereby mitigating DNA damage. Further research is certainly needed on the protective effects of DHA, including studies of the impact of repeated and prolonged use of DHA-containing products on skin cancer risk in humans.

A second study by the same group reported a negative impact of DHA. Peterson and colleagues found that DHA caused DNA damage when directly applied to cultured human keratinocyte cell lines [21]. Keratinocytes are generated in the deeper layers of the epidermis far below the stratum corneum. Whether DHA penetrates into the skin beyond the stratum corneum to affect the lower layers of the epidermis is not entirely clear [21]. One study measured radioactivity in human skin samples exposed to DHA to determine absorption and found that only about 0.5% of an applied dose of DHA was absorbed into the skin and was systemically available [22]. Further research is needed on the absorption and physiological impact of topically-applied DHA in the epidermis and systemically.

11.2.3 Inhalation and Exposure to Eyes and Mucous Membranes

The FDA has not reviewed the safety of DHA in the form of all-over sprays as delivered in spray booths because the appropriate safety data have not been reported by the industry. The FDA do not routinely review cosmetic products or ingredients per the Federal Food, Drug, and Cosmetic Act but will if safety data are provided to them [23]. The businesses that market cosmetics are ultimately responsible for assuring product safety and labeling, and local health departments are responsible for oversight of professionally-delivered products in salons and other businesses. However, the FDA has released an advisory that DHA should not be inhaled, ingested, or exposed to the entire area of the eyes and all mucous membranes, because little is known of the consequences of these types of exposures [24]. In one study of the sunless tanning safety practices of 17 tanning businesses in New York, Chicago, and Los Angeles, the vast majority of businesses reported to use safety precautions such as asking clients to close their eyes (100%), hold their breath (77%), and use sun protection (82%) [8]. Although so little data are available on safety practices of spray tanning businesses, some businesses minimize inhalation by conducting sessions in a well ventilated room and by using a filtered extraction fan which draws in the mist from the air.

11.2.4 Effect on Melanocytic Nevi

Two case studies reported changes in the appearance of melanocytic nevi (e.g., moles) resulting from sunless tanning use [25, 26]. In these cases, minute amounts of sunless tanning product residue were trapped in the contours of the surface of the nevi (rather than sloughing off as usual) and evident months following application. The clinical implication is that a benign lesion might seem to be darkening or otherwise changing in appearance as a result, which could prompt unnecessary excision, thus putting individuals at unnecessary risk. Sunless tanning users with melanocytic nevi should be aware of the potential for this phenomenon, and dermatologists should routinely assess use of sunless tanning products during skin examinations in their patients.

11.3 Clinical Uses

Although largely used cosmetically, some clinical uses of topically-applied DHA have been reported in the literature. For example, one report revealed that sunless tanning lotion applied over certain lesions generates a distinctive pattern in the skin that helps make the diagnosis of porokeratosis, a rare inherited skin disorder that can lead to the development of squamous cell cancer [27]. Several studies report the use of sunless tanning creams in camouflaging vitiligo in Caucasian, Asian, and Hispanic patients with this disorder [28–31]. Vitiligo is a skin disorder characterized by depigmentation of the skin, usually beginning on the hands, forearms, feet, and face, where skin and hair turn white in irregular patches. Patients experience significant distress from the alteration in physical appearance that results [32]. Sunless tanning applied to white areas has been shown to reduce the contrast between unaffected and affected skin, and most patients in one study reported moderate to high levels of satisfaction with the results [28].

11.4 Measurement

Self-report is the only means by which sunless tanning use has been assessed in the clinical research literature, however in animal studies, spectrophotometry has been reported [20]. In 2005, the National Cancer Institute (NCI) and Emory University co-sponsored a meeting of skin cancer prevention experts from academia, the Centers for Disease Control and Prevention (CDC), and the American Cancer Society (ACS) to develop a consensus about core survey items to measure self-reported skin cancer prevention behaviors among children and adults [33]. Consensus on the measurement of both indoor tanning and sunless tanning items was achieved and published in a report. Items were first taken from those used in published reports or in ongoing studies and then underwent both expert consensus discussion as well as participant cognitive testing. The final items recommended for measurement of sunless tanning in research are published elsewhere (see Ref. [33]). Objective measures of skin color (e.g., colorimetry, spectrophotometry) may be difficult to interpret in clinical research to the extent that participants expose themselves to both DHA and UVR, each of which independently contributes to skin color. Research is needed to examine how objective measures of skin color can be used under these circumstances.

11.5 Prevalence of Use

Although sunless tanning products are one of the fastest growing types of skin care products, a small minority of the population uses them. Five published studies have examined the prevalence of use of DHA-based sunless tanning products, four in the U.S. [19, 34–36] and two in Australia [37]. Rates of past year sunless tanning use in the U.S. have been reported from 11–46% depending on the nature and age of the sample. In the National Cancer Institute's Health Information National Trends Survey [38], which used a representative sample of U.S. adults, 11% of respondents reported past year use [35]. Among past-year users, 13% reported more than 25 uses, 12% reported 11–24 uses, 35% reported 3–10 uses, and 40% reported 1–2 uses. Women were three times more likely to report use of sunless tanners than men [35].

A U.S. study using a college student convenience sample observed much higher rates of use, with 23% reporting one or more past year use of sunless tanners with mean frequency of 11 uses [34]. No differences were found in sunless tanning use by gender [34]. Similarly, Brooks and colleagues observed past year use rates of 22% in a sample of young adults ages 18–30 years old recruited on the streets of Boston near universities and recreational areas [19]. However, gender differences were observed in that study, with 80% of self-applied lotion users and 64% of spray tan facility users being female. In a randomized trial sample of Boston-area female beach visitors, 46% of participants reported having used sunless tanners in the past year [36].

In Australia, reported rates of sunless tanning are somewhat lower than the U.S., but studies were also older. Two population-based studies (one in South Australia [39] and one in New South Wales [37]) both reported rates of past year use of sunless tanning of 8%, and one revealed lifetime rate of use at 22% [37]. However, among women aged 18–24, rates of past year use were 25%, which is consistent with use rates in this age group in the U.S. (e.g., [34]) To the extent that sunless tanning use is increasing over time, lower reported rates of sunless use in Australia versus the U.S. may be due to the fact that the Australian surveys were conducted in 1999 [39] and 2000 [37] while U.S. studies were based on surveys conducted in 2003 [34], 2004 [19], and 2005 [35] and 2006 [36]. More recent data are needed to examine whether rates of use are increasing and for which type of products (i.e., self-applied versus professionally applied).

11.6 Association of Sunless Tanning and Other UVR-Related Behaviors

The literature is mixed as to whether sunless tanning use is associated more strongly with high-risk tanning behaviors or UVR protective behaviors. Two studies reported higher rates of sunburns among users of sunless tanning products [19, 39], and one revealed higher rates of indoor tanning among users [19]. Higher rates of burns may be related to misconceptions about the degree of sun protection afforded by sunless tanners as well as the possibility that sunless tanner users are more likely to want a tan and to pursue multiple forms of tanning. On the other hand, studies have found that sunless tanner users have higher rates of sunscreen use than nonusers [35, 37, 39, 40]. This finding could possibly be suggestive of higher levels of sun exposure and therefore greater use of sunscreen and other safety behaviors among sunless tanning users relative to non-users. One study defined respondents as either "exclusive" sunless tanners (users who do not engage in indoor tanning) or "mixed" tanners (users of both sunless and indoor tanning) on sun protection behavior [35]. Among people using sunless tanners, 78% were "exclusive" sunless tanners while 12% were "mixed" tanners. Exclusive sunless tanners had higher rates of sunscreen compared to both indoor and non-tanners and higher rates of other forms of protective behavior (e.g., clothing) compared to indoor tanners [35]. These findings suggest that the majority of users of sunless tanners do not engage in indoor tanning, and users have greater sun protection behaviors relative to indoor tanners and non-tanners. Such users appear to represent a safety-conscious subgroup of sunless tanners who appear to desire a tan but actively protect their skin. Future research on the subtypes of sunless tanners might further clarify the association of sunless tanning with related risk and safety behaviors.

The extent to which sunless tanning offsets a previously existing tanning habit has only been explored in one study. Patrons of a tanning business for a sunless tanning visit were interviewed directly after their visit about any changes in their indoor tanning habits resulting from their sunless tanning [41]. Almost three quarters (73%) reported that they had decreased their indoor tanning frequency since they began sunless tanning, while only 7% reported having increased their indoor tanning frequency.

11.7 Promotion of Sunless Tanning for Skin Cancer Prevention

Because sunless tanning is a largely and relatively safe alternative to UVR tanning, the American Academy of Dermatology [10], Skin Cancer Foundation [11], and other organizations devoted to the prevention and control of skin cancer recommend that people who want to be tan use sunless tanners instead of the sun or indoor UVR sources. Two randomized trials have tested whether the promotion of sunless tanning within the context of a sun safety message influenced sun exposure and protective behavior [36, 42]. One trial randomized 146 college students and tested whether adding free sunless tanning samples to a sun safety intervention that included UV facial photography and a brief informational video on photoaging and sun protection practices improved sun protection practices compared to the same intervention without sunless tanning and a control condition [42]. Results revealed no added effect of free sunless tanner samples at the 1-month follow-up. However, no evidence of harm in promoting sunless tanning was observed, with mean levels of sun protection in the sunless plus intervention condition exceeding means in the intervention condition, albeit nonsignificantly. Non-significant findings may have resulted because participants reported low rates of tanning at baseline and very few actually used the sunless tanning sample that was provided, with many reporting no interest or negative past experiences with the color and odor of sunless tanners.

Building upon this research is a second randomized trial in which 250 female beach visitors were enrolled during their beach visit to participate in an intervention that included motivational messages to use sunless tanners as an alternative to UV tanning, instructions for proper use of sunless tanning products, attractive images of women with sunless tans, application of sunless tanning product and free take home sample, skin cancer education, and UV facial photography [43]. All participants in this study applied a small amount of sunless tanner on the top of their hand during the intervention session to experience the color firsthand and were given detailed application instructions. At 2 months, intervention participants reduced their sunbathing significantly more than those in a survey-only control group and reported significantly fewer sunburns and greater use of protective clothing [36]. At 1 year, intervention participants reported significant decreases in sunbathing and increases in sunless tanning relative to control participants, but no differences on sun protection or sunburns. Rates of sunless tanning increased by 9% among intervention participants at 2-months follow-up and by another 6% at 1-year follow-up. Of the intervention participants who tried sunless tanning for the first time by 2-months and 1-year following the intervention, the majority (64% and 75%, respectively) used it more than once, which strongly suggests product acceptance. In neither trial was there evidence for a negative impact of sunless tanning on risk behavior.

Additional research is needed to understand the characteristics of tanners most likely to adopt sunless tanning and how to facilitate a complete switch from UVR tanning to sunless tanning. Tanners whose UVR exposure is primarily motivated by the desire to be tan may be good targets for sunless tanning interventions, whereas tanners whose UVR exposure is motivated by social reasons, relaxation, and/or recreation might not be amenable to adopting sunless tanning or to changing their UVR exposure as a result of adopting sunless tanning.

11.8 Sunless Tanning in Development

New agents that create a tanned appearance without exposure to UVR are currently under development for cosmetic purposes, but more importantly for the clinical purpose of preventing skin cancer in people at high risk. One new class of agents operates by stimulating melanin production without UVR exposure [44]. Tans are the result of melanin production induced by UVR, and this increase in melanin in the skin serves to shield the skin from the harms of additional UVR exposure. However, UVR exposure also induces DNA damage, which offsets the advantage conferred by melanin. If melanin could be produced without UVR exposure, not only would a "real" tan be created, but it would then protect the skin from the damaging effects of UVR. This could be of great benefit to individuals who naturally have low levels of melanin and are at high risk for skin cancer (i.e., individuals with light skin tones). The products that are in development contain analogs of melanin-stimulating hormone (MSH), which stimulate the melanocortonin-1 receptor (MC1R), which increases melanin production in the skin, thereby creating a tan and enhancing DNA repair [44]. Both topical [45] and injectable [46] agents are currently being researched.

If both MC1R genes are damaged as is often the case in redheads, these MSH-analogs are ineffective because they operate by stimulating these receptors. For that reason, research is underway on agents that bypass the MC1R. Topically-applied forskolin, a substance that stimulates adenyl cyclase activity in the skin, has been shown to induce a tan that has a UV protection effect [47]. Also circumventing the MC1R pathway is an agent composed of DNA fragments called T-oligos, which have certain sequences that trigger tanning and DNA repair in the skin [44]. T-oligos has been shown to increase the melanin content in the skin, reduce markers of DNA damage following UVR exposure, and increase levels of a cancer-suppressor protein [48]. Research is ongoing to develop a "super" sunscreen that stimulates melanin production, creates a tan, protects the skin from sunburn, and buffers the skin against UV-induced DNA damage.

11.9 Promising Directions for Future Research on Sunless Tanning

Research on sunless tanning is currently in its infancy. Many questions are still unanswered, and promising new products that may have health benefits remain in early developmental phases. In terms of DHA-based sunless tanners, which are now ubiquitous in the marketplace, further research is needed to understand the characteristics of new users, frequent users, and exclusive users. As has been explored among indoor tanners [49], sunless tanning use subtypes should be identified to increase our understanding of the most common patterns of use. Relatedly, additional research is needed on the effect of adopting sunless tanning on other tanning behavior as well as protection behaviors. Very little is known about whether the use of sunless tanning results in changes in risk and safety habits. One study revealed evidence that adoption of sunless tanning offset an indoor tanning habit in the majority of tanners [41], which suggests a role for sunless tanning in skin cancer prevention efforts targeting indoor tanners specifically. The use of sunless tanning in the context of such skin cancer prevention efforts should be further explored. One weakness of sunless tanning as a behavioral alternative to outdoor and indoor tanning is that it lacks the relaxation, social, and recreational qualities that can be had with other forms of tanning. Tanners who are primarily motivated by appearance as opposed to relaxation/mood management, social, or recreational reasons, would seem to be the most receptive targets to sunless tanning interventions. Finally, future research on sunless tanning products that improve upon the limitations of DHA-based products is needed. Products currently in development that provide a significant source of UVR protection, produce an authentic tan by operating on melanin pathways, and impact DNA-repair mechanisms would have a much stronger clinical benefit than DHA-based products. However, the risks and adverse effects of triggering melanin production in this way would need to be fully explored.

11.10 Conclusions

DHA-based products as well as cosmetic bronzers are currently the only safe sunless tanning products on the market. About one in ten Americans report having used sunless tanners in the previous year [35], nearly ¼ of college students have done so, and nearly half of beach visitors [19, 34, 36]. Women compose the majority of users. Many misconceptions exist about sunless tanning products, including that they are not safer than the sun, or that they are a source of sun protection. Education is clearly needed on the proper use of sunless tanning products. DHA does not inherently provide a significant source of sun protection and users should be informed of the importance of engaging in sun protective practices during the use of such products. Research has revealed that promoting sunless tanning to tanners within the context of a skin cancer prevention public health message can be helpful in reducing UVR risk behavior and facilitating protective behaviors. Further research is also needed on the development of products that have potential for both cosmetic and skin health benefits. Encouraging UVR tanners (both indoor and outdoor) to switch to sunless tanning could have an important health impact, even though sunless tanning has been considered a cosmetic, more so than a healthcare tool. Healthcare providers could play an important role in encouraging patients who strongly desire a tan to consider switching to the safer alternative of sunless tanning, while engaging in optimal sun protection, to maximize skin health.

References

1. Draelos ZD (2002) Self-tanning lotions: Are they a healthy way to achieve a tan? Am J Clin Dermatol 3(5):317–318
2. Wittgenstein E, Berry HK (1960) Staining of skin with dihydroxyacetone. Science 132:894–895
3. Blau S, Kanof NB, Simonson L (1960) Dihydroxyacetone (DHA) – a keratin coloring agent. Arch Dermatol 85:501–504
4. Maibach HI, Kligman AM (1960) Dihydroxyacetone: a suntan simulating agent. Arch Dermatol 82:505–507
5. Levy SB (1992) Dihydroxyacetone-containing sunless or self-tanning lotions. J Am Acad Dermatol 27(6 Pt 1):989–993
6. United States Department of Health and Human Services (2002) Code of Federal Regulations, Title 21, Food and Drug Administration, Vol. 1, Part 73, Section 73. 1150. 2002.
7. White-Sax B (2004) Education key to growing sun care sales. *Drug Store News*, p 42
8. Fu JM, Dusza SW, Halpern AC (2004) Sunless tanning. J Am Acad Dermatol 50(5):706–713
9. US Food and Drug Administration (2005) Tanning pills. http://www.fda.gov/Cosmetics/ProductandIngredientSafety/ProductInformation/ucm134217.htm. Accessed 20 Apr 2010
10. American Academy of Dermatology (2010) Be sun smart. http://www.aad.org/public/sun/smart.html. Accessed 20 Apr 2010
11. Skin Cancer Foundation (2010) Tan in a can. http://www.skincancer.org/tan-in-a-can.html. Accessed 20 Apr 2010
12. American Cancer Society (2010) Can melanoma skin cancer be prevented? http://www.cancer.org/docroot/CRI/content/CRI_2_2_2X_Can_melanoma_skin_cancer_be_prevented_50.asp. Accessed 20 Apr 2010
13. Morren M, Dooms-Goossens A, Heidbuchel M et al (1991) Contact allergy to dihydroxyacetone. Contact Dermatitis 25(5):326–327
14. Kimura T (2009) Contact dermatitis caused by sunless tanning treatment with dihydroxyacetone in hairless descendants of Mexican hairless dogs. Environ Toxicol 24(5):506–512
15. Akin FJ, Marlowe E (1984) Non-carcinogenicity of dihydroxyacetone by skin painting. J Environ Pathol Toxicol Oncol 5(4–5):349–351
16. Fusaro RM, Rice EG (2005) The Maillard reaction for sunlight protection. Ann N Y Acad Sci 1043:174–183

17. Jung K, Seifert M, Herrling T et al (2008) UV-generated free radicals (FR) in skin: their prevention by sunscreens and their induction by self-tanning agents. Spectrochim Acta A Mol Biomol Spectrosc 69(5):1423–1428
18. US Food and Drug Administration (2000) Required warning statement for tanning products without sunscreen. http://www.fda.gov/Cosmetics/ProductandIngredientSafety/ProductInformation/ucm134252.htm. Accessed 20 Apr 2010
19. Brooks K, Brooks D, Dajani Z et al (2006) Use of artificial tanning products among young adults. J Am Acad Dermatol 54(6):1060–1066
20. Petersen AM, Takiya L, Finley R (2003) Meta-analysis of trials of interventions to improve medication adherence. Am J Health Syst Pharm 60:657–665
21. Petersen AB, Wulf HC, Gniadecki R et al (2004) Dihydroxyacetone, the active browning ingredient in sunless tanning lotions, induces DNA damage, cell-cycle block and apoptosis in cultured HaCaT keratinocytes. Mutat Res 560(2):173–186
22. Yourick JJ, Koenig ML, Yourick DL et al (2004) Fate of chemicals in skin after dermal application: Does the in vitro skin reservoir affect the estimate of systemic absorption? Toxicol Appl Pharmacol 195(3):309–320
23. US Food and Drug Administration (2010) Federal food, drug and cosmetic act (FD&C Act). http://www.fda.gov/RegulatoryInformation/Legislation/FederalFoodDrugandCosmeticActFDCAct/default.htm. Accessed 20 Apr 2010
24. US Food and Drug Administration (2009) Sunless tanners and bronzers. http://www.fda.gov/Cosmetics/ProductandIngredientSafety/ProductInformation/ucm134064.htm. Accessed 20 Apr 2010
25. Affleck AG, Perkins W (2008) Fake tan as a cause of a change in a pigmented lesion: seborrhoeic keratosis or melanocytic naevus. Clin Exp Dermatol 33(1):105–106
26. Martin JM, Monteagudo C, Calduch L et al (2007) Dermatoscopic changes in acquired melanocytic nevi and seborrheic keratoses after the application of a self-tanning airbrush. Arch Dermatol 143(11):1453–1455
27. Katugampola RP, Finlay AY (2006) Fake sun tan diagnosis of porokeratosis. J Eur Acad Dermatol Venereol 20(2):224–226
28. Rajatanavin N, Suwanachote S, Kulkollakarn S (2008) Dihydroxyacetone: a safe camouflaging option in vitiligo. Int J Dermatol 47(4):402–406
29. Hsu S (2008) Camouflaging vitiligo with dihydroxyacetone. Dermatol Online J 14(8):23
30. Suga A, Ikejima S, Matsuba H et al (2002) Medical pearl: DHA application for camouflaging segmental vitiligo and piebald lesions. J Am Acad Dermatol 47:436–438
31. Fesq H, Brockow K, Strom K et al (2001) Dihydroxyacetone in a new formulation – a powerful therapeutic option in vitiligo. Dermatology 203(3):241–243
32. Kent G, Al'Abadie M (1996) Psychologic effects of vitiligo: a critical incident analysis. J Am Acad Dermatol 35(6):895–898
33. Lazovich D, Forster J (2005) Indoor tanning by adolescents: prevalence, practices and policies. Eur J Cancer 41(1):20–27
34. Mosher CE, Danoff-Burg S (2005) Social predictors of sunscreen and self-tanning product use. J Am Coll Health 54(3):166–168
35. Stryker JE, Yaroch AL, Moser RP et al (2007) Prevalence of sunless tanning product use and related behaviors among adults in the United States: results from a national survey. J Am Acad Dermatol 56(3):387–390
36. Pagoto SL, Schneider KL, Oleski J et al (2010) The sunless study: a beach-randomized trial of a skin cancer prevention intervention featuring sunless tanning. Arch Dermatol 146:979–984
37. Girgis A, Tzelepis F, Paul CL et al (2003) Australians' use of fake tanning lotions: another piece of the puzzle. Aust N Z J Public Health 27(5):529–532
38. Cantor D, Covell J, Davis TC, et al. (2005) Health information national trends survey (HINTS 2005): Final report 2005. http://hints.cancer.gov/docs/HINTS_2005_Final_Report.pdf. Accessed 20 Apr 2010
39. Beckmann KR, Kirke BA, McCaul KA et al (2001) Use of fake tanning lotions in the South Australian population. Med J Aust 174(2):75–78

40. Dixson H, Cappiello M, Borland R (1997) Reaction to the 1994/1995 SunSmart campaign: results from a representative household survey of Victorians. SunSmart Evaluation Studies 5. Anti-Cancer Council of Victoria, Melbourne, p 64
41. Sheehan DJ, Lesher JL Jr (2005) The effect of sunless tanning on behavior in the sun: a pilot study. South Med J 98(12):1192–1195
42. Mahler HI, Kulik JA, Harrell J et al (2005) Effects of UV photographs, photoaging information, and use of sunless tanning lotion on sun protection behaviors. Arch Dermatol 141(3):373–380
43. Pagoto SL, Schneider KL, Oleski J et al (2009) Design and methods for a cluster randomized trial of the sunless study: a skin cancer prevention intervention promoting sunless tanning among beach visitors. BMC Public Health 9:50
44. Wickelgren I (2007) Skin biology. A healthy tan? Science 315(5816):1214–1216
45. Abdel-Malek ZA, Kadekaro AL, Kavanagh RJ et al (2006) Melanoma prevention strategy based on using tetrapeptide alpha-MSH analogs that protect human melanocytes from UV-induced DNA damage and cytotoxicity. FASEB J 20(9):1561–1563
46. Barnetson RS, Ooi TK, Zhuang L et al (2006) [Nle4-D-Phe7]-alpha-melanocyte-stimulating hormone significantly increased pigmentation and decreased UV damage in fair-skinned Caucasian volunteers. J Invest Dermatol 126(8):1869–1878
47. Spry ML, Vanover JC, Scott T et al (2009) Prolonged treatment of fair-skinned mice with topical forskolin causes persistent tanning and UV protection. Pigment Cell Melanoma Res 22(2):219–229
48. Arad S, Konnikov N, Goukassian DA et al (2006) T-oligos augment UV-induced protective responses in human skin. FASEB J 20(11):1895–1897
49. Hillhouse J, Turrisi R, Shields AL (2007) Patterns of indoor tanning use: implications for clinical interventions. Arch Dermatol 143(12):1530–1535

Chapter 12
International Perspectives on Indoor Tanning

Jennifer Hay and Samara Lipsky

Abstract This chapter reviews the English language literature regarding the prevalence, frequency, and predictors of sunbed use outside of the United States. The literature examining sunbed use across European and non-European countries indicates prevalent use, especially in young women. In Sweden an extensive body of literature indicates that over 50% of females in their teens and twenties have used sunbeds at least once. In other European countries, the literature is less extensive but shows that rates of sunbed use are consistent with those reported in Sweden, at least in Northern Europe. There is evidence of a North–South gradient with lower prevalence of use in Southern Europe. The United Kingdom, Canada, and Australia document more modest rates of sunbed use. Of concern is the relatively high rate of heavy sunbed use in many countries. While measurement strategies differ, 20–25% of sunbed users indicate very frequent sunbed use recently, or over their lifetime. The predictors of sunbed use are consistent across studies and include female gender, younger age, higher socioeconomic status and urban locale, beliefs that a tan will improve appearance and that a "base tan" is needed before going on vacation, lower skin sensitivity to sunburn, heightened tobacco use and outdoor tanning behavior. The belief that tanned skin is attractive is a motivation for tanning, especially in young women, and needs to be confronted as new mass media and individual-level interventions are developed.

J. Hay (✉)
Department of Psychiatry & Behavioral Sciences, Memorial Sloan-Kettering Cancer Center, New York, USA
e-mail: hayj@mskcc.org

S. Lipsky
Department of Psychiatry & Behavioral Sciences, Memorial Sloan-Kettering Cancer Center, New York, USA
e-mail: samaralipsky@alum.emory.edu

Department of Psychology, Yeshiva University, New York, USA
e-mail: lipskys@mskcc.org

C.J. Heckman and S.L. Manne (eds.), *Shedding Light on Indoor Tanning*,
DOI 10.1007/978-94-007-2048-0_12, © Springer Science+Business Media B.V. 2012

Keywords Sunbed use • Indoor tanning • International perspectives • Solaria • Prevalence • Frequency • Predictors • Tanning beliefs • Knowledge • Health behaviors • Skin sensitivity • Skin cancer family history • Regulation • Jennifer Hay • Samara Lipsky

Abbreviations

UVR Ultraviolet radiation
UK United Kingdom

12.1 Introduction

Since the 1960s and 1970s indoor commercial tanning facilities (also called tanning salons or solaria) and sunbeds (also called sunlamps or tanning booths), as well as home tanning equipment, have become widely available in many countries around the world. Given the developing consensus that indoor tanning increases risk of developing cancers, including both cutaneous and ocular melanoma [1–7], the World Health Organization has recommended that people under age 18 should avoid the use of sunbeds entirely [8].

Despite these warnings, the use of sunbeds continues to proliferate internationally. In this chapter, we review the English language literature regarding the prevalence, frequency, and predictors of sunbed use outside of the United States. We also discuss the extent of sunbed regulation and users' adherence with recommended safety guidelines. Interestingly, we did not find literature documenting commercial or home indoor tanning use in Asian countries; this may be because high value is placed on lighter rather than darker skin in China [9] and Southeast Asia [10]. We close with a summary and series of recommendations for future research and intervention to address continuing sunbed use internationally.

12.2 Prevalence in Europe

In the 1980s, the use of sunbeds became increasingly common across Western Europe. Northern European countries adopted sunbed use earlier, with use extending to Southern countries over a period of about 10 years [11]. In Germany, sunbeds were frequently used by individuals as a medical treatment for skin conditions, with physicians providing a medical prescription for use in the 1980s and earlier. The use of sunbeds in Belgium, France, and Germany has increased quickly, with reasons for tanning shifting from medical purposes to aesthetic purposes over the 1980s [1].

In Germany and Belgium, home-use sunbeds have been quite common [1]. In many Northern European countries, sunbeds are often advertised as a treatment for seasonal affective disorder, or "winter depression" [11–14], even though white fluorescent light that does not contain ultraviolet radiation (UVR) is a standard effective treatment for this disorder [15]. A recent web-based survey of 8,178 general population volunteers across 12 countries compared rates of intentional tanning (indoor and outdoor tanning were not determined separately) in European compared to non-European countries (including Australia, Israel, and the United States), and found the highest rates of intentional tanning were reported in Sweden and Italy and the lowest rates in the United Kingdom (UK) and Poland [16]. In the next sections, we discuss in more detail the prevalence and frequency of indoor tanning reported for Western European countries.

In Sweden, the inconsistent levels of daily sun – even in the summer – likely lead those who are motivated to achieve tanned skin (mostly for appearance reasons) to pursue indoor tanning, or to travel south for sunny vacations. Sunbeds were first available in Sweden starting in 1978. By the early 1990s, sunbed use had become very common in Sweden with up to 70% of adolescents and young adults reporting that they had ever used one, with rates tending to be higher in urban areas [15, 17] and among teenagers rather than adults. The most frequently reported venues for indoor tanning in Sweden include gyms, tanning salons, and public indoor swimming pools [18]. Boldeman and colleagues reported that in the early 1990s the most marked increase in use by age in Swedish teenagers is between 8th and 9th grade, between age 14 and 15, in parallel with heightened tobacco use at this age [19]. This may also coincide with the beginning of high school. Teenagers who were the least satisfied with themselves (concerning height, weight, looks, hair, complexion, body and personality) used sunbeds more frequently (14%). A 1994 study in Southern Sweden found that 31% of women and 16% of men over the age of 30 reported that they had ever used sunbeds, compared to 56% of women, and 12% of men age 30 and under who had ever used sunbeds [20]. Furthermore, a population-based study conducted in non-urban areas in the mid 1990s reported that 16% of girls and 4% of boys of three age groups (ages 13, 15 and 17) had ever used sunbeds [21]. Rates may be higher in university populations. A study published in 1999 conducted by Jerkegren and colleagues [22] surveyed 304 Swedish university students (on average, age 24) in 1995 and found that most (86% of women and 73% of men) had ever used a sunbed; 12% overall had used a sunbed more than ten times during the last year. More recently, a 2001 study of Swedish adults found that 44% of 1,752 randomly-selected individuals ages 18–37 from the general population had ever used sunbeds [23].

The overall prevalence of sunbed use has been decreasing in Sweden through the 1990s, perhaps due to Swedish media attention to skin cancer, sun protection, and the risks associated with sunbed use that was widely disseminated by the late 1990s; yet, the proportion of those engaging in heavy sunbed use may actually be increasing. Boldeman and colleagues conducted two cross-sectional surveys that examined trends in sunbed use (defined as use ≥4 times in the past year) in urban Stockholm County teenagers (ages 15–19). The authors found that sunbed use among girls had dropped by half through the 1990s, from approximately 70% to 45% of teenage girls surveyed from 1993 to 1999, and from approximately 44% to 19% for boys [24]. However, rates of high-frequency use of sunbeds (≥10 sessions/year) actually increased from 1993 to 1997, from 14% to 17% among girls, and from 9% to 22% among boys [25]. Boldeman and colleagues also documented significant levels of long-term use, with 20% of female users ages 30–39 and 10% of males in this age group reporting that they had used sunbeds for 10 years or more [18]. The only existing Swedish educational intervention consisted of one school-based session, and was designed to impact indoor and outdoor tanning (targeting 13–15 year-old Swedish adolescents); however, it was not found to be efficacious in increasing intentions to abstain from tanning or in decreasing beliefs about the attractiveness of tanned skin, let alone indoor tanning behavior itself [26]. While further intervention work is certainly warranted, the lack of effect in this study may be due to the one-session nature of the intervention, as well as the social and cultural resilience of attitudes regarding the attractiveness of tanned skin.

In other Northern European countries, studies document rates of sunbed use that are comparable to those reported in Sweden. One 1993 study examined a large sample of high school students in Norway, and found that among 15,169 students drawn from 103 high schools, about three-quarters (75%) of girls had used a sunbed during the past 12 months; 17% had used one 15 times or more. Among boys, 65% had used a sunbed in the past 12 months; 6% had used one 15 times or more. Of those who reported that they had used sunbeds in the past year, girls did so on average two to three times, and boys about two times [27]. In a 2002 study of home sunbed users in Holland [28], a telephone survey of 349 individuals (age range from 18 to 82 years; mean age = 41) showed that 39% used their home sunbeds year-round, on average 5–6 times per month.

In Denmark, sunbed use has been increasing over the past 15 years. From 1994 to 2004, sunbed use (reported in the past 2 years) jumped from 35% to 50% [29]. Most recently, a population-based sample of Danish teenagers and adults (N = 3,427) indicated that 29% had used a sunbed in the past 12 months, and these rates were higher (59%) for teenage girls, but also high for boys (41%); modal age of first use was 14–15 years old as in Sweden [30].

In contrast, use of sunbeds is significantly lower in the UK, but there is great variation in use across geographical regions within the UK. Across a number of studies, ever use is documented at about 7% among adults [31], teenagers [32], and even young children (ages 8–11; [33]). However, sunbed use in 11–17 year olds is reported to be more common in Northwest England (Liverpool 20%) than in Southeast England (Southhampton 6.2%; [32]). Another recent study found much higher rates (43% ever used sunbeds) among teenagers in Merseyside in Northwest England [34]. Finally, a study of 14 and 15 year-olds in South Wales found that 32% of girls and 10% of boys had previously used a sunbed, with almost 6% of girls reporting using a sunbed every week [35]. One study from Scotland of sunbed users found that half the sample who were sunbed users began use between the ages of 16 and 24, and the primary reasons for sunbed use in the UK were: to feel good (25%), to treat a medical condition such as acne, eczema, or psoriasis (17%), to get a tan (16%), to look good (14%), and to tan before holiday [36].

Recently, Thomson and colleagues examined large population based-samples of teenagers from six targeted "cities" in England and found that on average 6% of 11–17 year-olds had ever used sunbeds. However, sunbed use in 15–17 year olds is more common (11.2%) than in 11–14 year olds (1.8%). Previous sunbed use was reported in 8.6% of girls and 3.5% of boys and higher rates were found in the North than in the South with the lowest rates in London (3.2%). Of those who used a sunbed, 38.4% had used one at least weekly and more boys than girls used a sunbed once a year or less. Modal age of first use was 14 for both genders in all cities except for Sunderland where it was 15; 7% of sunbed users started using sunbeds before age 12. Settings where indoor tanning was performed included tanning at an outlet with staff supervision followed by using a sunbed at home. The primary reasons for not using a sunbed in England were being uninterested (64% of boys; 44% of girls), and a health risk (33% of boys; 51% of girls). Other less common reasons were: expense, lack of access and lack of parental consent to participate. The more frequent use of sunbeds in Northern areas could be due to the availability of sunbeds or merely a shift from outdoor to indoor tanning given the reduced opportunity for outdoor tanning in the North, where it is colder [37].

Despite the lower overall use of sunbeds in the UK, heavy use appears to be as common as that reported in Swedish populations. In a Scottish sample of sunbed users recruited at indoor tanning facilities (84% female; majority aged 25–34), 26% of sunbed users reported 50 or more sessions in the past 12 months [36]. McGinley and colleagues surveyed 205 sunbed users (majority female; 73% under age 35) and found that 16% reported having had over 100 indoor tanning sessions in a year [38].

In Germany, two recent studies have found that between 29% and 47% of German residents report that they have used a sunbed at least once in their lives [39, 40]. One survey of 500 Mannheim, Germany residents (ages 18–45) found that 21% had used a sunbed in the past year. About a third (28.4%) of sunbed users had done so before their 18th birthday, but there was a high proportion of users over age 35, which may be unique to a German population. One additional recent study of 1,242 study volunteers recruited during a skin cancer screening program in North Rhine-Westphalia Germany found that 15% of the sample had used sunbeds at least ten times in the past year, with reasons reported for tanning including a desire for a tan (68%) and preparation for sunny holidays (59%). About half (56%) reported that they had experienced sunburns following the use of sunbeds [40]. German settings where indoor tanning was performed included solaria (73%), followed by sports facilities such as fitness studios (15%), swimming baths (15%), and saunas (8%; [41]).

In France, a large (N=7,303) study of individuals across the country found that 15% of adults surveyed reported indoor tanning from 1994 to 1995, with indoor tanning being significantly more common in the North (13%) than in the South (10%). Primary reasons for indoor tanning in France were for "aesthetic purposes" (30%) or to achieve a "pre-vacation tan" (17%; [42]).

There is little literature documenting the patterns of sunbed use in more Southern European countries. Sunbed use was examined in 764 young people (ages 16–21) selected by stratified random sampling by age from ten high schools and a medical school in Naples, Italy. About 12% reported using sunbeds [43].

12.3 Prevalence in Canada and Australia

Canada and Australia document more modest rates of sunbed use compared to most European countries. In Canada, approximately one-fifth of adults report sunbed use. A 1996 population-based telephone survey of indoor tanning behaviors in 1,003 adults (ages 18–60) in Montreal, Quebec found that 20% had visited a commercial tanning salon at least once over the past 5 years; 11% had used one in the past 12 months. Most users were women aged 18–34. Of those who said they had used a tanning salon in the past year, most (77%) intended to do so again. Perhaps of most concern, 26% had experienced an adverse health event associated with their use of tanning salons, most often skin burns, but the experience of an adverse health event did not reduce intentions to return for future use. Reasons for use included the desire to improve appearance (stated more by women than men) and to acquire a pre-vacation tan (stated more by men than women; [44]).

Rates of sunbed use appear to be approximately 10–15% among Canadian teenagers. In 2006, the Canadian Cancer Society reported that 12% of grade 9–10 students reported tanning salon use in the past year [45]. Similar rates have been reported more recently [46] in a school-based survey of 1,200 grade 10 students in the Thames Valley, Ontario. Most users (75%) were female, and there was a strong normative element to teenagers' indoor tanning, as teens were more likely to tan if their parents, siblings, and friends did so. The survey found that 14% of grade 10 students had used an indoor tanning facility in the past year, and of those who had used one, 26% had used these facilities at least once per week during the last year. Attitudes highly related to indoor tanning included beliefs in the improved appearance of tanned skin. In Canada, home-use sunbeds are quite common, as in Germany and Belgium [1].

Rates of use in Australia are reported to be much lower compared to most European countries or Canada, which may be due to the preponderance of sunny warm weather and individuals with moderately UVR-sensitive skin, as well as the relative ubiquity of mass media skin cancer prevention campaigns in Australia. A 2006 study of 9,298 adults (ages 20–75) in Australia found that 8.8% of adults in Australia reported ever using a sunbed, but less than 1% of respondents had done so in the past year. Frequency of solarium use in Australia showed that in 2003–2004, the majority of sunbed users reported that their pattern of use was consistent with concentrated bursts (27.7%) or one or two sessions (34.9%) while approximately one-third of sunbed users reported more frequent use (weekly or two or more times a week). Settings where indoor tanning was performed included tanning at the hairdresser/beautician (49.4%) followed by solaria (30.1%). Fewer respondents (13.3%) tanned at the gym [47].

Recently, Francis and colleagues surveyed a large population based-sample of 11,509 adolescents (12–17) and adults (18–69) across two time periods (2003/2004 and 2006/2007) in Australia. In 2006/2007 on average 10.6% of 18–69 year olds and 2.5% of younger adolescents had ever used a sunbed. Sunbed use was reported to be more common in women ages 18–24 (17.1%) and 25–44 (20.7%), with 6.7% of women ages 18–24 having used a sunbed in the past year. Frequency of solarium use in Australia showed that over the last 12 months, the majority of sunbed users reported 5–24 sessions (50%), while 15% of sunbed users reported more frequent use (25 or more times). The primary reasons for using a sunbed in Australia were: "preparing for a special event or holiday" (30.6%), "vanity/to get a tan/I'm too pale" (19.9%), and "fading tan" (15.2%). Although overall sunbed use in Australia is low compared to other countries, it remains highest among young adult women, as in other countries [48].

12.4 Predictors of Sunbed Use

In the international literature on sunbed use, the predictors of high utilization are very consistent across studies and countries. In this section, we discuss predictors of sunbed use from a number of domains, including demographics, tanning beliefs, skin sensitivity and skin cancer family history, knowledge of the risks of sunbed use, and the concurrent adoption of other health risk behaviors.

12.4.1 Demographics

Universally, females are more likely to have used sunbeds at least once, and to report more frequent use than males. These gender differences have been documented across every country that has examined sunbed use, including all European countries as well as in Canada and Australia. One Swedish study found that sunbed use was higher in girls who were less satisfied with themselves or perceived themselves as less attractive but among boys who were most satisfied with themselves [19, 21]. Although levels of use were difficult to compare across studies due to the variation of ages measured in the sample size, highest levels of use have been documented from early teens to mid 30, with age of first use on average being 14–15 years old [16, 19, 27, 30, 38, 42, 47, 49].

Furthermore, higher socioeconomic status, employment, and urban residence have been related to more prevalent and frequent sunbed use across a number of countries. In Sweden, urban populations tend to report more frequent use than rural populations [19, 21]. In Germany and Italy, sunbed use is related to higher socioeconomic status and being employed [39, 40]. In Australia as well, employment status, higher income, and urban residence predict higher utilization of sunbeds [47]. In contrast, in the UK, children from lower socioeconomic backgrounds have been reported to use sunbeds more often than those from higher socioeconomic backgrounds [32]. While no reasons for this finding are proposed by the authors, it is possible that those from lower socioeconomic backgrounds in the UK, both children and their parents, may be less aware of the dangers of sunbed use, and that this may deter use in those from higher socioeconomic groups in the UK.

12.4.2 *Tanning Motivations*

Motivations to tan include beliefs that one looks good with a tan, and beliefs that a pre-vacation or "base tan" is desirable (to look tan on the first day of vacation and to prepare the skin for a sun), all appear to be quite important across many countries [38, 42]. For example, a German study found that most sunbed users did so in order to tan (68%) and in preparation for sunny holidays (59%; [40]). In Canada, correlates of indoor tanning include the desire to improve appearance and to acquire a pre-vacation tan [44]. In the UK, primary reasons reported for sunbed use are to "feel good" (25%), treat medical conditions (17%), "get a tan" (16%), "look good" (14%), and "tan before holiday" (13%; [34, 36, 38]), with similar reasons reported in France [42]. In Italy, sunbed users tended to value sunbeds in order to look better and to engage in sunbathing as well [43].

12.4.3 *Skin Sensitivity and Skin Cancer Family History*

Overall, sunbed users tend to be those who are less sensitive to sunburn [31, 39, 43, 44]; nevertheless, many sunbed users report that they have experienced erythema (sunburns). In Sweden, almost half (44%) of sunbed users have reported erythema during use [16, 17]. French sunbed users tend to have more sun-sensitive phenotypes than non-sunbed users, and to report freckling, lighter hair, and fairer complexions. Of those who reported indoor tanning in France, 57% had Fitzpatrick skin type I (always burns, never tans), 45% had skin type II (always burns, sometimes tans), 36% had skin type III (sometimes burns, always tans), and 25% had skin type IV (never burns, always tans; [42]). Sixty-five percent of those who frequently used sunbeds in the UK reported having no moles on their skin compared with 35% who have moles (e.g. $p<0.001$; [49]), indicating the possibility that those with moles may be aware that this is a skin cancer risk factor. This possibility was not assessed in this study, however. Also in the UK, family and personal history of skin cancer are related to reduced utilization of sunbeds. In one UK study, ever use of sunbeds was doubled (48% versus 24%) in those without, versus with, a family history of skin cancer [49]. In summary, sunbed users tend to have fewer genetic risk factors for skin cancer, such as being less sensitive to sunburn, having fewer moles and not having a personal family history of the disease.

12.4.4 Knowledge of the Risks of Sunbed Use

Interestingly, some studies show that sunbed users are more aware of the harms of use than are non-users. For example, high levels of knowledge of the risks associated with indoor and outdoor tanning have been reported in Sweden [17, 19] and in Germany [41]. Boldeman and colleagues reported that sunbed users were well-informed about the dangers of UVR exposure and were receptive to learning more about it, but that sunbed users who acknowledge sunbed dangers in general may not feel that the risks apply to them personally [19]. A large 2004 study of over 2,600 Swedish teenagers (ages 13, 15, and 17) found higher levels of accurate knowledge and greater risk awareness among indoor tanners as compared to non-tanners. At the same time, this study found higher rates of unrealistic optimism among tanners as compared to non-tanners, so that tanners believed that the general risks of tanning, and of developing skin cancer, did not apply to them personally. Awareness of risks, indoor tanning behavior, and unrealistic optimism were all particularly high for teenage girls [50]. There have been mixed findings in Italy concerning knowledge of sunbed use. High levels of knowledge concerning the harms of UVR to skin have been reported (90%) in young people aged 16–21. However, Italian sunbed users in this study tended to value sunbeds in order to look better and to disagree that sunbeds had harmful effects [43].

12.4.5 Health Behaviors

In terms of health behaviors, sunbed use is consistently associated with outdoor tanning behaviors [19, 51] as well as with heightened tobacco use, but also with increased participation in physical activities, such as aerobics or strength training [39]. Sunbed users in a large French study reported more frequent outdoor tanning than non-sunbed users [42]. In the UK, two studies have documented that significantly more smokers (61%) reported using a sunbed compared with nonsmokers [34, 49], and the same finding has been reported in France [42] and Sweden [19]. The relationship of sunbed use with participation in physical activities may be explained by the fact that, in Germany and other countries, sunbeds are often available in sports facilities including gyms and swimming pools [41]. In Australia as well, sunbeds are located in hairdressers, "solaria," as well as gyms [47]. The availability of tanning facilities in exercise or health contexts may convey an implicit endorsement of the health benefits of tanning in these countries, or could be related to a desire to maintain a tan that was achieved through outdoor physical activities or that would emphasize muscle definition, but these specific motivations have not been documented in this literature.

12.5 Regulation of Sunbed Use

Some regulation of the international sunbed industry began nearly 20 years ago, with the first Swedish National regulation of the UVR emission properties of sunbeds introduced in 1991 [19]. While some countries have developed specific rules for sunbed installation, operation, and use, there is no standardized regulation across the European Union. In fact in some countries, such as Germany, a greater percentage of sunlamp exposure is from home equipment, for which there is no enforcement of usage guidelines [2]. In the Netherlands, UK, and Canada such recommendations are developed by, or with, the sunbed industry itself [11, 52]. However, some Canadian Provinces, such as New Brunswick and Nunavut in Northern Canada have prohibited indoor tanning among those under age 18 [53], consistent with WHO recommendations [8]. In 2009, a law was passed in Brazil banning all UVR cosmetic tanning [54].

In 2006, a law was passed in Santiago, Chile requiring that all users of commercial sunbed facilities must be made fully aware of the risks involved in use. A study of compliance with this law utilized a questionnaire of the policies of 24 tanning salons in the Santiago area, conducted by telephone by researchers posing as potential clients. Some procedures did change in response to the new law, for example, spontaneous information about risks was given by 46% of salons after law enactment versus 25% before, and goggle use increased to 25% after law enactment versus 12% before. However, both before and after the law, salon staff did not provide accurate procedural or risk information, for instance, both before and after the law was enacted, some tanning salon staff told potential clients that tanning was "beneficial for children." Even after enactment of the law, only 50% required written parental consent for minors to engage in indoor tanning, despite the law's requirement for such documentation. Written educational information was not provided before or after law enactment, and only two salons required signature of informed consent by adult clients after enactment of the law, which was required by law [53]. We could find no other literature documenting sunbed use in Chile. In summary, the enactment of such laws are a step in the right direction but is not adequate to assure that sunbed staff are well-informed or that they adhere to the full intentions of such regulations.

While there has been a start at regulating the indoor tanning industry, poor communication between the salons and their patrons regarding the recommended use of sunbeds has been documented widely. Sunbed users across a number of countries report that they received little to no instruction on avoiding sunbed sunburns or how to protect their eyes through the use of goggles. For example, in Italy [43], about one-fifth (23%) of indoor tanners had suffered a sunburn from sunbeds, but of those, only 60% discontinued use after a sunburn. Two recent German studies found that about half (46–56%) of indoor tanners reported that they had experienced sunburns following the use of sunbeds and that very few (16%) used goggles to protect their eyes [40, 41]. Rates of sunbed erythema (defined by the survey as reddened skin after sunbed use in 1993; burn with smarting redness in 1999) was common and noted by nearly half of Swedish sunbed users in 1993 (45% of female users; 43% of male users) and was lower but remained common in 1999 (29% and 19% of female and male users, respectively; [24]). One-third (29.7%) of Canadian and Scottish sunbed users ([38, 44]) reported that they never or hardly ever used goggles while in sunbeds. An interesting study conducted in Poland [55] found that tanning salons where staff were better informed concerning the UVR emitted by their units were more likely to recommend the usage of goggles and sunscreen, but while 73% of salons surveyed provided goggles, only 18% required sunbed users to wear them during tanning sessions; another study conducted in Buenos Aires, Argentina [3] had similar findings.

A Holland telephone survey of 349 sunbed users revealed that most (94%) thought that they used their sunbeds in a safe manner, yet this figure dropped to 37% when "safe" use was described to them (using eye protection, cleaning the skin before use, slowly increasing exposure time, and never exceeding maximum exposure). Self-efficacy for safe use, and anticipated regret about unsafe use, were important predictors of intentions to use home sunbeds "safely" in the future in this study [28].

Accordingly, there has been a start at sunbed regulation that has mainly comprised country-specific guidelines, yet there is little indication in the literature that there are enforcement strategies in place or that guidelines for less risky use are consistently communicated to sunbed users or the general public. While the WHO recommendation that individuals under the age of 18 avoid the use of sunbeds entirely [8] is a welcome international message, there is no research we could identify establishing the extent of adherence with this recommendation subsequent to 2005. Given the continued high levels of use among teenagers, it is likely that the regulation has not yet had the desired impact on sunbed use in this population.

12.6 Summary and Directions for Future Research

The literature examining sunbed use across European and non-European countries indicates quite prevalent use especially among, but not limited to, young women. The well-developed literature in Sweden documents that over 50% of Swedish girls in their teens and twenties have used sunbeds at least once. In other European countries, the empirical literature is less extensive but shows that rates of sunbed use are consistent with those reported in Sweden, at least in Northern European countries. Indeed, there is evidence of a North–South gradient with use diminishing somewhat in more Southern European countries. Of note, both the UK and Australia document relatively more modest rates of indoor tanning use than European countries. In Sweden there is some indication that sunbed use may be on the decline.

There are many areas in need of further research. First, outside of Sweden, there are few studies documenting sunbed use in each country, and comparison across countries is hampered by the use of non-comparable measurement strategies, sampling frames, and populations assessed. Research documenting trends in commercial and home sunbed use across countries using similar methods to monitor prevalence of use across time is a priority. Such research would allow the possibility of reliably tracking the effect of new sunbed regulations, as well as the effect of public health messages emphasizing the risks associated with sunbed use that will likely increase as the harms of use become better understood. Additionally, the use of comparable methods would allow for more precise specification of the probable North–South gradient in use, as well as differential motivations for sunbed use in Northern versus more Southern countries. For example, is the more frequent use in Sweden and more Northern countries due to differences in attitudes towards tanning, more availability of sunbeds, or merely a shift from outdoor to indoor tanning given the reduced opportunity for outdoor tanning due to climate? Furthermore, there is almost a total lack of published intervention work outside of the US. Given the relatively high rates of sunbed use across many countries, the development of innovative population-based interventions to prevent such use is a top priority. Given that young people are likely at the greatest long-term risk of the negative repercussions of sunbed use and the clear indication that modal age of first use is 14–15 years, population-based interventions targeted to this age group would likely be a good place to start. Both mass media and individual-level interventions might be useful modalities to reduce sunbed use. These efforts could capitalize on successful country-specific interventions developed to address outdoor tanning behavior [23].

Of perhaps most concern, however, is the relatively high rate of heavy use of sunbeds documented in many countries. Again, while measurement strategies differ, about 20–25% of sunbed users may be identified as heavy users as evidenced by their frequent recent or lifetime use or the fact that they have consistently used sunbeds over long periods of time; these rates may be increasing, at least in Sweden. Again, the research in each country is unfortunately quite limited, so the rates of heavy use are approximate and require confirmation. Additionally, intervention approaches with these heavy users may need to be distinct from approaches used for prevention among the general public. Research examining tanning attitudes and reasons for their heavy use have not focused on these heavy users in non-United States populations but could help to determine the unique factors dictating habitual or repeated use of sunbeds. Recently, the addictive nature of sunbed use has been examined [56–58] and requires research internationally to examine potential cultural and ethnic differences in addiction to indoor tanning (See Chap. 7 by Shah and colleagues). Additionally, the diverse contexts where sunbeds are used, including home and gymnasia or swimming pools, may lead to diverse practice patterns that remain unspecified at this time. Such research could lead to the development of novel, relevant, and efficacious interventions to limit or stop frequent sunbed use throughout the world.

Despite the differences in use rates, the predictors of sunbed initiation and use are quite consistent across studies and include female gender, younger age among adults, higher socioeconomic status and urban locale, beliefs that one looks good with a tan and that one needs a "base tan" before going on vacation (for either skin protection or for appearance reasons), less skin sensitivity to sunburn, as well as outdoor tanning behavior and heightened tobacco use. Less work has examined the covariates of heavy use. These factors will be quite useful in targeting interventions to those cohorts at a higher risk of sunbed use initiation and maintenance, as well as providing essential content for new intervention efforts. Indeed, the belief that tanned skin is attractive is a primary motivation for tanning, especially in young women, and needs to be confronted head-on as new interventions are developed. In contrast, accurate knowledge and higher risk perceptions are either not associated with sunbed use or are positively related to it, indicating that knowledge of the risks of sunbeds is well-understood among users and that these beliefs have not served as deterrents to use. We caution against the development of interventions, either for the general population or higher-risk groups that focus exclusively on the risks of tanning, as there is consistent evidence that knowledge of these risks is not associated with reduced sunbed use internationally. In contrast, interventions that target specific at-risk subpopulations, address appearance motives, and suggest alternative, pleasurable, health-focused activities may be useful in addressing increasing sunbed use in many countries.

Overall, there is generally a limited amount of country-specific data on indoor tanning across the world. Further research, including developing and testing potential interventions, is an important priority to specify and address the proliferation of this important health risk behavior.

References

1. Autier P, Dore JF, Lejeune F et al (1994) Cutaneous malignant melanoma and exposure to sunlamps or sunbeds: an EORTC multicenter case-control study in Belgium, France and Germany. EORTC Melanoma Cooperative Group. Int J Cancer 58(6):809–813
2. Autier P, Joarlette M, Lejeune F et al (1991) Cutaneous malignant melanoma and exposure to sunlamps and sunbeds: a descriptive study in Belgium. Melanoma Res 1(1):69–74
3. Chouela E, Pellerano G, Bessone A et al (1999) Sunbed use in Buenos Aires, Argentina. Photodermatol Photoimmunol Photomed 15(3–4):100–103
4. Diffey BL (1987) Analysis of the risk of skin cancer from sunlight and solaria in subjects living in northern Europe. Photodermatology 4:118–126
5. El Ghissassi F, Baan R, Straif K et al (2009) A review of human carcinogens – part D: radiation. Lancet Oncol 10(8):751–752
6. International Agency for Research on Cancer (2007) The association of use of sunbeds with cutaneous malignant melanoma and other skin cancers: a systematic review. Int J Cancer 120(5):1116–1122
7. Walter SD, Marrett LD, From L et al (1990) The association of cutaneous malignant melanoma with the use of sunbeds and sunlamps. Am J Epidemiol 131(2):232–243
8. World Health Organization (2005) The World Health Organization recommends that under 18s should not use a sunbed. J Adv Nurs 51(6):662
9. Dixson BJ, Dixson AF, Li B et al (2007) Studies of human physique and sexual attractiveness: sexual preferences of men and women in China. Am J Hum Biol 19(1):88–95
10. Sahay S, Piran N (1997) Skin-color preferences and body satisfaction among South Asian-Canadian and European-Canadian female university students. J Soc Psychol 137(2):161–171
11. Autier P (2004) Perspectives in melanoma prevention: the case of sunbeds. Eur J Cancer 40(16):2367–2376
12. Howland RH (2009) An overview of seasonal affective disorder and its treatment options. Phys Sportsmed 37(4):104–115

13. Prasko J (2008) Bright light therapy. Neuro Endocrinol Lett 29(Suppl 1):33–64
14. Shirani A, St Louis EK (2009) Illuminating rationale and uses for light therapy. J Clin Sleep Med 5(2):155–163
15. Lam RW, Buchanan A, Mador JA et al (1992) The effects of ultraviolet-A wavelengths in light therapy for seasonal depression. J Affect Disord 24(4):237–243
16. Branstrom R, Chang YM, Kasparian N et al (2010) Melanoma risk factors, perceived threat and intentional tanning: an international online survey. Eur J Cancer Prev 19(3):216–226
17. Boldeman C, Beitner H, Jansson B et al (1996) Sunbed use in relation to phenotype, erythema, sunscreen use and skin diseases. A questionnaire survey among Swedish adolescents. Br J Dermatol 135(5):712–716
18. Boldeman C, Branstrom R, Dal H et al (2001) Tanning habits and sunburn in a Swedish population age 13–50 years. Eur J Cancer 37(18):2441–2448
19. Boldeman C, Jansson B, Nilsson B et al (1997) Sunbed use in Swedish urban adolescents related to behavioral characteristics. Prev Med 26(1):114–119
20. Westerdahl J, Olsson H, Masback A et al (1994) Use of sunbeds or sunlamps and malignant melanoma in southern Sweden. Am J Epidemiol 140(8):691–699
21. Brandberg Y, Ullen H, Sjoberg L et al (1998) Sunbathing and sunbed use related to self-image in a randomized sample of Swedish adolescents. Eur J Cancer Prev 7(4):321–329
22. Jerkegren E, Sandrieser L, Brandberg Y et al (1999) Sun-related behaviour and melanoma awareness among Swedish university students. Eur J Cancer Prev 8(1):27–34
23. Branstrom R, Ullen H, Brandberg Y (2004) Attitudes, subjective norms and perception of behavioural control as predictors of sun-related behaviour in Swedish adults. Prev Med 39(5):992–999
24. Boldeman C, Jansson B, Dal H et al (2003) Sunbed use among Swedish adolescents in the 1990s: a decline with an unchanged relationship to health risk behaviors. Scand J Public Health 31(3):233–237
25. World Health Organization (1995) Protection against exposure to ultraviolet radiation. Geneva: WHO/EHC/UNEP
26. Kristjansson S, Helgason AR, Mansson-Brahme E et al (2003) 'You and your skin': a short-duration presentation of skin cancer prevention for teenagers. Health Educ Res 18(1):88–97
27. Wichstrom L (1994) Predictors of Norwegian adolescents' sunbathing and use of sunscreen. Health Psychol 13(5):412–420
28. Lechner L, De Vries H (2002) Sunbed use at home: risk behaviour and psychosocial determinants. Eur J Cancer Prev 11(4):333–341
29. Engholm G, Ferlay J, Christensen N et al (2008) NORDCAN: cancer incidence, mortality and prevalence in the Nordic countries, version 3.2. http://www.ancr.nu
30. Køster B, Thorgaard C, Clemmensen IH et al (2009) Sunbed use in the Danish population in 2007: a cross-sectional study. Prev Med 48(3):288–290
31. Diffey BL (2003) A quantitative estimate of melanoma mortality from ultraviolet A sunbed use in the U.K. Br J Dermatol 149(3):578–581
32. Canadian Cancer Society (2009) Sunbeds. Cancer Research UK Policy Statement. October 2009
33. Hamlet N, Kennedy K (2004) Reconnaissance study of sunbed use by primary school children in Lanarkshire. J Public Health (Oxf) 26(1):31–33
34. Mackay H, Lowe D, Edwards DJ et al (2007) A survey of 14–16 year olds as to their attitude toward and use of sunbeds. Health Educ J 66:141–152
35. Roberts DL, Foley K (2009) Sunbed use in children-time for new legislation? Br J Dermatol 161(1):193–194
36. Oliver H, Ferguson J, Moseley H (2007) Quantitative risk assessment of sunbeds: impact of new high power lamps. Br J Dermatol 157(2):350–356
37. Thomson CS, Woolnough S, Wickenden M et al (2010) Sunbed use in children aged 11–17 in England: face to face quota sampling surveys in the National Prevalence Study and Six Cities Study. BMJ 340:c877
38. McGinley J, Martin CJ, MacKie RM (1998) Sunbeds in current use in Scotland: a survey of their output and patterns of use. Br J Dermatol 139(3):428–438

39. Diehl K, Litaker DG, Greinert R et al (2010) The prevalence of current sunbed use and user characteristics: the SUN-Study 2008. Int J Public Health 55(5):513–516
40. Dissel M, Rotterdam S, Altmeyer P et al (2009) Indoor tanning in North Rhine-Westphalia Germany: a self-reported survey. Photodermatol Photoimmunol Photomed 25(2):94–100
41. Schneider S, Zimmerman S, Diehl K et al (2009) Sunbed use in German adults: risk awareness does not correlate with behaviour. Acta Derm Venereol 89(5):470–475
42. Ezzedine K, Malvy D, Mauger E et al (2008) Artificial and natural ultraviolet radiation exposure: beliefs and behaviour of 7200 French adults. J Eur Acad Dermatol Venereol 22(2):186–194
43. Monfrecola G, Fabbrocini G, Posteraro G et al (2000) What do young people think about the dangers of sunbathing, skin cancer and sunbeds? A questionnaire survey among Italians. Photodermatol Photoimmunol Photomed 16(1):15–18
44. Rhainds M, De Guire L, Claveau J (1999) A population-based survey on the use of artificial tanning devices in the Province of Quebec, Canada. J Am Acad Dermatol 40(4):572–576
45. Canadian Cancer Society (2006) Sun bed usage and attitudes among students in Ontario grades 7–12 youthography–summary results
46. Gordon D, Guenther L (2009) Tanning behavior of London-area youth. J Cutan Med Surg 13(1):22–32
47. Lawler SP, Kvaskoff M, DiSipio T et al (2006) Solaria use in Queensland, Australia. Aust N Z J Public Health 30(5):479–482
48. Francis K, Dobbinson S, Wakefield M et al (2010) Solarium use in Australia, recent trends and context. Aust N Z J Public Health 34(4):427–430
49. Amir Z, Wright A, Kernohan EE et al (2000) Attitudes, beliefs and behaviour regarding the use of sunbeds amongst healthcare workers in Bradford. Eur J Cancer Care (Engl) 9(2):76–79
50. Sjöberg L, Holm LE, Ullen H et al (2004) Tanning and risk perception in adolescents. Health Risk Soc 6(1):81–94
51. Brandberg Y, Sjoden PO, Rosdahl I (1997) Assessment of sun-related behaviour in individuals with dysplastic naevus syndrome: a comparison between diary recordings and questionnaire responses. Melanoma Res 7(4):347–351
52. Autier P (2005) Cutaneous malignant melanoma: facts about sunbeds and sunscreen. Expert Rev Anticancer Ther 5(5):821–833
53. Salomone C, Majerson D, Molgo M et al (2009) Tanning salons in Santiago, Chile: the knowledge of the staff in charge and the quality of information provided to potential clients before and after a new regulatory law. Photodermatol Photoimmunol Photomed 25(2):86–89
54. Teich M (2009) Nations unite against tanning: the impact of the IARC report. Skin Cancer Foundation August, 2009
55. Szepietowski JC, Nowicka D, Soter K et al (2002) Tanning salons in southwest Poland: a survey of safety standards and professional knowledge of the staff. Photodermatol Photoimmunol Photomed 18(4):179–182
56. Heckman CJ, Egleston BL, Wilson DB et al (2008) A preliminary investigation of the predictors of tanning dependence. Am J Health Behav 32(5):451–464
57. Mosher CE, Danoff-Burg S (2010) Addiction to indoor tanning: relation to anxiety, depression, and substance use. Arch Dermatol 146(4):412–417
58. Zeller S, Lazovich D, Forster J et al (2006) Do adolescent indoor tanners exhibit dependency? J Am Acad Dermatol 54(4):589–596

Chapter 13
Indoor Tanning: Past, Present, and Future

Sharon L. Manne and Carolyn J. Heckman

Abstract In this chapter, the authors summarize the information presented in the previous twelve chapters of this volume. The key content of each chapter is reviewed, as well as important areas where future research may be indicated.

Keywords Conclusions • Future research • Indoor tanning

Despite being a known carcinogen (see Chap. 6), tanning beds remain mostly unregulated because they are categorized as cosmetic devices by the Food and Drug Administration (FDA) in the United States. As a result of this lack of regulation, the indoor tanning industry has ballooned into a five billion dollar a year enterprise, and the US has witnessed an alarming increase in the incidence of skin cancer over the past 30 years. There are more than 30 million indoor tanning customers in the US and more than a million people use tanning beds each day [1, 2]. There are 25,000 free-standing indoor tanning salons in the US and an additional 20,000 outlets in health clubs and spas [3]. It is both impressive and frightening that there are more tanning facilities per square mile than McDonald's restaurants or Starbuck's coffee shops in major cities in the US [4]. Although rates of indoor tanning are low and possibly declining in some areas such as Australia (see Chap. 12), rates of indoor tanning in the past year have been difficult to quantify, particularly among teenagers

S.L. Manne (✉)
The Cancer Institute of New Jersey, UMDNJ-Robert Wood Johnson Medical School, 195 Little Albany Street, New Brunswick, NJ 08901, USA
e-mail: mannesl@umdnj.edu

C.J. Heckman
Cancer Prevention and Control Program, Fox Chase Cancer Center, 333 Cottman Avenue, Philadelphia, PA 19111, USA
e-mail: carolyn.heckman@fccc.edu

C.J. Heckman and S.L. Manne (eds.), *Shedding Light on Indoor Tanning*,
DOI 10.1007/978-94-007-2048-0_13, © Springer Science+Business Media B.V. 2012

and young adults in the US. Prevalence estimates range from 11% to 60% (see Chap. 3), with about 30% of young indoor tanners reporting they would have difficulty quitting [5].

How did a widely-recognized public health hazard become big business with a growing number of possibly-addicted consumers? In this volume, we have elucidated important historical and cultural, individual, business, societal, and regulation influences contributing to the popularity and accessibility of tanning beds. As outlined in Chap. 2, the emphasis on the perceived health benefits of sun exposure and the attractiveness of a tanned appearance has a long history dating back at least to the early 1900s and the industrial revolution. Over the course of the several decades, a cultural paradigm shift has occurred in which a tanned appearance has taken on greater social significance. The indoor tanning industry has both contributed to and benefitted from this cultural shift. In their efforts to increase their revenues, the industry has promulgated a number of false arguments about the benefits of tanning, has adopted an aggressive marketing approach to young consumers, and has not educated the tanning facility operators regarding the dangers of tanning or enforced existing regulations. Particularly damaging claims made by the tanning industry include the claim that indoor tanning is a good source of vitamin D, that tanning bed use does not cause cancer if done as intended, and that it is possible to acquire a "safe" tan (see Chaps. 8 and 10). This advertising approach has seemingly convinced the general public that tanning has fewer health risks than the facts suggest. It is evident that the indoor tanning industry's approach to marketing their devices, as well as the initial state and federal approach to regulating the industry, is very similar to the tobacco industry's approach in the twentieth century.

Given the challenges presented by the aggressive marketing efforts of the tanning industry and the lack of a comprehensive approach to addressing the utilization of tanning devices, what can be done to control the growth of the tanning industry and reduce indoor tanning, particularly among adolescents and young adults? Federal regulation measures to reduce rates of indoor tanning, such as restricting direct advertising and accessibility to youth and enacting an excise tax on tanning bed use, are approaches that have been used effectively to control the use of tobacco products. Some of these measures have been initiated already, but the US and other countries are probably years away from a comprehensive tanning industry regulation policy. The development of more appealing sunless tanning products may reduce tanning bed use, but it is possible that they will be used concurrently by a subset of frequent tanners or that they will be used as sunscreen, although they are not sunscreen agents (see Chap. 11). Therefore, sunless tanning may not ultimately curb tanning bed use among tanners who enjoy the mood-lifting and social effects of tanning in addition to its appearance enhancement effects.

Even with regulatory and alternative product-development efforts in place, similar to tobacco, indoor tanning is likely to remain a problem for a subset of the population. There are a number of psychological, biological, and social factors motivating tanning bed use that need to be better understood (see Chap. 4). Knowledge about the dangers of indoor tanning is high, and therefore knowledge-based interventions delivered alone will not prove efficacious. We know that young women are more frequent users of tanning beds and that appearance enhancement is a key motivating factor for indoor tanning. Based on this information, the limited number of extant intervention studies have targeted college-aged women and have adopted a largely appearance-based approach.

However, much less is understood about how appearance motivations interact with other factors such as social and media influences. Even less is known about what biological and psychological factors lead to tanning dependence (see Chap. 7). A subset of regular tanners use tanning as a method of affect regulation, and it is possible that affective factors play a stronger role in maintenance of the behavior than is presently appreciated. Like any harmful and possibly addictive health practice, tanning is a complex phenomenon. Factors that motivate the first visit to a tanning salon (e.g., response to targeted marketing, getting a tan before one goes on vacation or to the prom, accompanying a parent or friend) will differ from factors motivating a second and third visit, and a more complex set of factors likely contribute to sustained tanning behaviors. It is probable that compared to indoor tanning initiation, maintenance of tanning involves a fuller set of psychological, genetic, social, and biological vulnerabilities. Just as is the case with adolescents or adults who take their first alcoholic drink, those individuals who go on to become occasional drinkers differ in multiple regards compared to those who become dependent on alcohol. A greater theoretical and empirical understanding of tanning motivations and outcomes will inform the development of more effective interventions (see Chap. 9). There is very limited research on motivations for tanning, and even more limited intervention research. A more comprehensive framework for understanding indoor tanning from a cultural, societal, and bio-psychological perspective is necessary. Such efforts will enhance both behavioral interventions and public health efforts to reduce the use of indoor tanning.

In this volume, we have sought to present an overview of the history, medical risks, and motivators for indoor tanning. Behavioral research and public policy efforts in this area are in their infancy. With increased research, public education campaigns, as well as advocacy and regulation, it is our hope that rates of indoor tanning will be on the decline in the near future.

References

1. Kwon HT, Mayer JA, Walker KK et al (2003) Promotion of frequent tanning sessions by indoor tanning facilities: two studies. J Am Acad Dermatol 46:700–705
2. Spencer JM, Amonette RA (1995) Indoor tanning: Risks, benefits, and future trends. J Am Acad Dermatol 33:288–298
3. Cokkinides VE, Weinstock MA, O'Connell MC et al (2002) Use of indoor tanning sunlamps by US youth, ages 11–18 years, and by their parent or guardian caregivers: prevalence and correlates. Pediatrics 109(6):1124–1130
4. Hoerster KD, Garrow RL, Mayer JA et al (2009) Density of indoor tanning facilities in 116 large U.S. cities. Am J Prev Med 36(3):243–246
5. Zeller S, Lazovich D, Forster J et al (2006) Do adolescent indoor tanners exhibit dependency? J Am Acad Dermatol 54(4):589–596

Index

C.J. Heckman and S.L. Manne (eds.), *Shedding Light on Indoor Tanning*,
DOI 10.1007/978-94-007-2048-0, © Springer Science+Business Media B.V. 2012

W

Y